THE ISOMORPHISM
PROBLEM IN
COXETER GROUPS

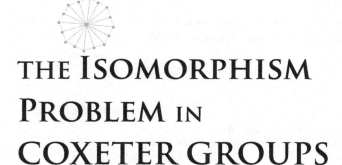

THE ISOMORPHISM PROBLEM IN COXETER GROUPS

Patrick Bahls

University of Illinois, Urbana-Champaign, USA

Imperial College Press

Published by

Imperial College Press
57 Shelton Street
Covent Garden
London WC2H 9HE

Distributed by

World Scientific Publishing Co. Pte. Ltd.
5 Toh Tuck Link, Singapore 596224
USA office: 27 Warren Street, Suite 401-402, Hackensack, NJ 07601
UK office: 57 Shelton Street, Covent Garden, London WC2H 9HE

British Library Cataloguing-in-Publication Data
A catalogue record for this book is available from the British Library.

ISBN-13 978-1-86094-554-0
ISBN-10 1-86094-554-6

Printed in Singapore

For Maggie

Preface

The outline of this volume is based roughly upon a series of lectures given to the Group Theory Seminar in the Department of Mathematics at the University of Illinois, Urbana-Champaign, in the fall of 2002. The text's content is drawn from the robust literature on questions related to the Isomorphism Problem for Coxeter groups. The material contained in that literature has been augmented with a sizable amount of preliminary material in order to make the text somewhat self-contained. With this supplementary material, the text should be accessible to any graduate student (or mature undergraduate student) who has taken courses in linear algebra and group theory. (For instance, basic knowledge of group theory, including group presentations and group actions is assumed.) Following is an outline of the text.

After a brief sketch of the historical development of the study of Coxeter groups, the first chapter provides an introduction to the general theory of these groups. A number of exercises are included in order to allow the student who is less familiar with Coxeter groups the opportunity to explore their nature. Chapter 1 should give the reader the background needed to approach the topics covered later in the text. Moreover, the first chapter highlights the interplay between the combinatorial and geometric characterizations of Coxeter groups in order to demonstrate to the reader the usefulness of both of these points of view.

The second chapter builds upon the first by surveying specific properties of Coxeter groups of importance later in the text. It is in this chapter that we compute the normalizers and centralizers of important subgroups of a given Coxeter group. Here we also discuss decompositions of Coxeter groups as free products with amalgamation, and we describe a few homological properties of Coxeter groups. A brief survey of the Word and Conjugacy Problems is provided as well. Much of this information will be used in later chapters.

The third chapter begins by outlining various characterizations of uniqueness in Coxeter groups, including the concept of rigidity. The questions with which the remainder of the text concerns itself are then stated. This chapter contains few proofs, as here we seek merely to provide an overview of the results regarding rigidity while indicating the ways in which they are related to one another.

The next four chapters fill in the gaps in rigor contained in Chapter 3. Reasonably complete proofs of primary results are provided (the reader is often asked to supply details in the exercises). Roughly speaking, we address the primary results in such a manner as to separate those whose proofs require arguments of a geometric flavor from those whose proofs are more combinatorial. However, as the reader will notice, it is often difficult to separate results in this fashion.

In the final chapter we address questions related to the Isomorphism Problem. Here, for instance, we investigate the structure of $\text{Aut}(w)$ for a given Coxeter group W. Chapter 7 also considers the concept of rigidity in the setting of Artin groups.

This volume is in no way meant to be a comprehensive treatment of Coxeter groups. The reader should be aware that these groups are of considerable importance in a number of fields not addressed in this book, including low-dimensional topology, Lie theory, representation theory, and algebraic geometry.

As of the writing of this text, the Isomorphism Problem for Coxeter groups (the "Holy Grail" of the study of rigidity) remains unsolved. However, it is possible, if not likely, that a solution to this problem will be discovered quite soon. Such a solution may or may not make use of methods similar to those examined in this volume. In any case, a number of the partial results that we consider here may be rendered obsolete. Why, then, consider this text?

It is my hope that the reader will take from the text not only an understanding of the questions involved in approaching the Isomorphism Problem, but also an appreciation for the various methods with which such questions are addressed. Indeed, these methods highlight the structure and beauty of the groups themselves, and provide direction for investigation into properties not considered here. It is for this reason, if for no other, that so many different viewpoints are adopted in the course of our study. Each of these viewpoints has its own merits, and each leads to methods and machinery that are applicable in the solution of many other problems in the theory of Coxeter groups.

P. Bahls, Urbana, 2004

Contents

Chapter 1

Preliminaries on Coxeter groups

In this chapter we develop much of the basic theory regarding Coxeter groups that will be used throughout the remainder of the text. Although the material contained in this chapter is not comprehensive, enough background is provided to make study of the text largely self-contained. The interested reader may consult either [Bourbaki (1981)] or [Humphreys (1990)] for a more exhaustive treatment of Coxeter groups in general. Additional references are provided as needed in order to direct the reader to more throrough discussions of other topics mentioned in this chapter.

1.1   Historical background and motivation

Coxeter groups have been studied *pro se* for nearly three quarters of a century. One of the first rigorous treatments of these groups is found in the work of H.S.M. Coxeter ([Coxeter (1934)]), although a number of talented mathematicians of the late nineteenth and early twentieth centuries considered such groups at some point in their work (often without explicit indication of the relevant group structure).

Coxeter groups arose as a natural generalization of groups of symmetry and crystallographic groups. For Coxeter, at least initially, they are the groups generated by involutions R_i, in which the angles at which the reflecting "mirrors" corresponding to the involutions intersect are prescribed by means of relations of the form

$$(R_i R_j)^{m_{ij}}$$

for specified integers m_{ij}. (Coxeter goes so far as to invite the reader to create a fundamental domain for the action of a finite Coxeter group by using actual mirrors and a candle flame!)

Throughout his life, Coxeter retained the term he used for these groups in his earliest investigations, calling the objects of his study *groups generated by reflections* or *reflection groups*. (They would not be called *Coxeter groups* until Tits coined this term many years later.) Much of the basic theory of these groups is due to Coxeter himself. For instance, the enumeration of all finite Coxeter groups was first provided by Coxeter in one of the earliest papers written on the subject ([Coxeter (1935)]).

Following the lead established by Coxeter's initial studies, Coxeter groups have frequently been considered from a geometric point of view. Much as was done by Coxeter, Coxeter groups are often defined as groups generated by reflections in a particular vector space which is assigned the appropriate bilinear form (see 1.2.2).

More recent geometric characterizations see Coxeter groups acting isometrically on spaces which exhibit very nice metric properties. For example, we will consider below (in 1.5.3) a complex defined by M. Davis ([Davis (1983)]), building on the work of E. B. Vinberg. This complex, which carries a CAT(0) metric (as proven by Moussong in [Moussong (1996)]), will prove very useful in our studies. It is closely related to another complex, introduced by Tits. The latter object (generally called the Coxeter complex) admits significant generalization, leading to the very rich theory of buildings. In Chapter 6 we consider yet another complex (the chamber complex, also related to buildings) on which the given group acts. Coxeter groups have also been shown to act on trees ([Dranishnikov and Januszkiewicz (1999)]) and on CAT(0) cubical complexes ([Niblo and Reeves (2003)]).

Coxeter groups can also be realized in a straightforward combinatorial fashion, wherein the presentation of the group is of central importance, and analysis of the group is frequently performed with little reference to the group's geometric structure. Methods such as small cancellation theory and van Kampen diagrams can often be applied. The combinatorial viewpoint will facilitate the proofs of many results, and it is the manner in which we first address Coxeter groups in the following sections.

In whatever light they are studied, Coxeter groups arise naturally in crystallography, Lie theory, commutative algebra, representation theory, low-dimensional topology, combinatorics, and geometric group theory. They are useful creatures indeed!

It is clear that the subject of Coxeter groups is a broad one, any study of these groups must by necessity adopt a narrowed focus. We ask, then: what sort of problems will we address in this volume?

Our primary concern will be to understand the ways in which Coxeter groups may be presented. For instance, we may ask to what extent presentations for these groups are "unique". The ultimate goal is to find a solution to the Isomorphism Problem for Coxeter groups. That is, can we define an algorithm which tells us, given two presentations for Coxeter groups, if the presentations define isomorphic groups? This is the third, and most difficult, of the fundamental problems of combinatorial group theory posed by Max Dehn in his seminal 1910 work on surface groups. (The other two, the Word Problem and the Conjugacy Problem, are both known to be solvable for Coxeter groups. See Section 2.4.) Although a solution to the Isomorphism Problem has not yet been obtained, a number of the results proven in this text provide close approximations to a solution in a number of special cases.

The Isomorphism Problem is closely related to the structure of the automorphism group $Aut(W)$. We will examine $Aut(W)$ more carefully in the final chapter of this volume.

We now begin our study by defining Coxeter groups, any by putting forth the rudiments of the general theory of these groups that will be useful later on.

1.2 Coxeter systems and Coxeter groups

1.2.1 *A combinatorial definition*

A *Coxeter system* is a pair (W, S) for which $S = \{s_i \mid i \in I\}$ is a distinguished generating set with index set I, and

$$W \cong \langle S \mid R \rangle, \text{ for } R = \{(s_i s_j)^{m_{ij}} \mid m_{ij} \in \{1, 2, ..., \infty\}\}. \quad (1.1)$$

We demand that $m_{ij} = m_{ji}$ and $m_{ij} = 1 \Leftrightarrow i = j$ for all $i, j \in I$. By $m_{ij} = \infty$ we mean that $s_i s_j$ has infinite order. (Such infinite relators may be omitted. We include them in order to remain consistent with the definition frequently adopted, in which the values m_{ij} are related to the entries of the symmetric *Coxeter matrix*, defined in 1.2.2 below.) It can be proven (see [Bourbaki (1981)]) that the product $s_i s_j$ has order m_{ij} in the group W so defined. (This is not true *a priori*. In fact, it is not even clear that the relations R do not identify some distinct generators s_i and s_j.)

The index set I may have any cardinality, though we will often consider only the case in which I is finite. The set S is known as the *fundamental generating set* for the system (W, S). If W is a group for which there exists a presentation as in (1.1), we call W a *Coxeter group*.

There is a convenient way of representing Coxeter groups by means of graphs. We first review the necessary definitions from graph theory, as the notation and terminology are far from standard.

A *graph* X is a pair (V, E), where V is the set of *vertices* of X, and E is the set of *edges* of X. We will be concerned only with undirected graphs for which two vertices are connected by at most one edge, and for which the endpoints of any edge are distinct. Thus for our purposes an edge is completely determined by its two endpoints, and we may identify any element $e \in E$ with a pair $\{v_1, v_2\}$, $v_1, v_2 \in V$. We will denote the edge so defined by $[v_1 v_2]$. We say that $v_1, v_2 \in V$ are *adjacent* if $e = [v_1 v_2]$ is an edge in E; in this case, both v_1 and v_2 are said to be *incident* the edge e. Given $v_0, v_1, ..., v_k \in V$ such that $[v_i v_{i+1}] \in E$ for all i, $0 \le i \le n-1$, the sequence $v_0, v_1, ..., v_k$ defines a *path* P (*of length* n) in X. The graph X is said to be *connected* if for any two vertices $v_1, v_2 \in V$, there is a path from v_1 to v_2. An *edge-labeled graph* is a graph which comes equipped with a function lab : $E \to A$ assigning a label lab(e) from the set A to each edge $e \in E$.

Now let (W, S) be a Coxeter system. We define the *Coxeter diagram* $\mathcal{V}$ corresponding to (W, S) to be an edge-labeled graph whose vertex set is in one-to-one correspondence with S, and for which there is an edge $[s_i s_j]$ in $\mathcal{V}$ whenever $m_{ij} < \infty$. For any edge $e = [s_i s_j]$ in $\mathcal{V}$, we define lab(e) $= m_{ij}$. We often suppress the labeling function itself and simply say that an edge e is labeled by a particular number.

It is clear that given $\mathcal{V}$, the presentation $\langle S \mid R \rangle$ is completely determined, and *vice versa*.

Remark 1.1   The Coxeter diagram must be distinguished from the *Coxeter graph*, used in [Bourbaki (1981)] and [Humphreys (1990)], among other writings. In the latter construction, no edge is included between generators s_i and s_j which commute, and an edge is included between generators s_i and s_j for which $s_i s_j$ has infinite order. Furthermore, the label 3 is omitted from any edge $[s_i s_j]$ for which $s_i s_j$ has order 3. This second construction (which facilitates study of direct product decompositions, rather than free product decompositions) is often more convenient when there is a great deal of commutativity between elements of S. In order to maintain consistency throughout the text, we will use only Coxeter diagrams as defined above. Unfortunately, this may require a good deal of mental "translation" between graphs and diagrams for those more familiar with the former!

We will frequently omit the word "Coxeter" when referring to Coxeter groups, systems, and diagrams, as long as there will be no confusion.

There are some classes of Coxeter systems to which we will devote a great deal of attention. Whether or not a given Coxeter group belongs to a certain class can often be decided merely by examining its diagram. Therefore we define some of these classes now in terms of the Coxeter diagram.

Let (W, S) be a Coxeter system with diagram $\mathcal{V}$. If every edge in $\mathcal{V}$ has label 2, we say that (W, S) (or $\mathcal{V}$) is *right-angled*. At the other extreme, we say that (W, S) (or $\mathcal{V}$) is *large-type* if no edge in $\mathcal{V}$ is labeled 2. (Such systems are also called *skew-angled*, as in [Mühlherr and Weidmann (2002)].) Appel and Schupp (in [Appell and Schupp (1983)]) introduce a yet more extreme case, saying a Coxeter system is *extra-large-type* if every edge in the corresponding diagram has label at least 4. If every edge in $\mathcal{V}$ is even, we call (W, S) (or $\mathcal{V}$) *even*. If, given the group W, a system (W, S) exists which is right angled, (extra-)large-type, or even, we call W itself right angled, (extra-)large-type, or even, respectively. (However, we will see that the distinction between systems and groups made here will not always be necessary. This fact is the focus of much of this volume.)

One more definition for now. It may be that W can be easily decomposed as a direct product in a nontrivial fashion. If $S = S_1 \cup S_2$ such that $S_1 \cap S_2 = \emptyset$ and $s_i \in S_i (i = 1, 2) \Rightarrow s_1 s_2 = s_2 s_1$, then $W = W_1 \times W_2$, where $W_i = \langle S_i \rangle$. If this sort of decomposition is not possible, we call W *irreducible*. Many questions about Coxeter groups can be reduced to questions about irreducible groups.

We will have need to examine still other classes of groups as we continue our study.

1.2.2 *A geometric definition*

Suppose we are given an index set I and a collection of values $m_{ij} \in \{1, 2, ..., \infty\}$ $(i, j \in I)$ satisfying $m_{ij} = m_{ji}$ and $m_{ij} = 1 \Leftrightarrow i = j$. (As before, I can be either finite or infinite, although we will often be concerned only with finite I.) This information is often encoded in a symmetric matrix (called the *Coxeter matrix*) $A = (a_{ij})$ for which $a_{ij} = -\cos \frac{\pi}{m_{ij}}$ when $m_{ij} < \infty$ and $a_{ij} \leq -1$ when $m_{ij} = \infty$. This method of bookkeeping suggests the definition which now follows.

Let V be a vector space of dimension $|I|$ over the real numbers, with basis $\{\alpha_i \mid i \in I\}$ index by I. We define a symmetric bilinear form $(\cdot, \cdot)$ on

V by means of the matrix A. That is, we demand that

$$(\alpha_i, \alpha_j) = a_{ij}, \tag{1.2}$$

Denote by H_i the subspace of V orthogonal to α_i relative to this bilinear form; H_i is then complementary to the line in V containing α_i.

We note that the bilinear form $(\cdot, \cdot)$ motivates the terminology *right-angled* and *skew-angled* introduced in 1.2.1. Indeed, if W is right-angled and α_i and α_j are any two distinct vectors defined as above, either α_i and α_j are orthogonal or $(\alpha_i, \alpha_j) \leq -1$. In case W is skew-angled, no two vectors α_i and α_j are orthogonal to one another.

We now define the *reflection* $r_i : V \to V$ corresponding to α_i by

$$r_i(v) = v - 2(\alpha_i, v)\alpha_i \tag{1.3}$$

for every $v \in V$. It is easy to see that each r_i preserves the bilinear form, that each point of H_i remains fixed by r_i, and that $r_i(\alpha_i) = -\alpha_i$ (see exercises), as shown in Figure 1.1. The mappings r_i (with the operation of composition) generate a subgroup W of the group $\mathrm{GL}(V)$ of linear transformations of V. This group is a Coxeter group as defined in 1.2.1, with system $(W, \{r_i \mid i \in I\})$.

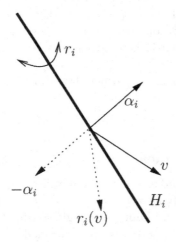

Fig. 1.1   The vector α_i and its corresponding reflection

Remark 1.2   Alternately, we could begin with W as in 1.2.1 and map each element $s_i \in S$ to $r_i \in \mathrm{GL}(V)$ for V a vector space of dimension $|S|$

spanned by α_i, where the bilinear form is defined as in (1.2), and r_i is defined as in (1.3). That this gives a faithful linear representation of W is proven carefully in [Humphreys (1990)]. That is, Coxeter groups are linear groups, and therefore enjoy a number of desirable properties. (For instance, all linear groups have solvable word problem, are residually finite, and are virtually torsion free.)

The vectors α_i are called *simple roots*. Denote by Φ the set $\{w(\alpha_i) \mid w \in W, i \in I\}$. (Here of course $w(\alpha_i)$ is the vector which results from the application of the linear transformation w to α_i.) Then Φ is known as the *root system* corresponding to W. Each element of Φ is called a *root*. The reader is asked to show in Exercise 1 that Φ is finite if and only if W is finite.

Because every α_i is a unit vector, and because every w preserves the bilinear form $(\cdot, \cdot)$, every element of Φ is a unit vector. Because $\{\alpha_i \mid i \in I\}$ is a basis of V, every root α can be expressed as a unique linear combination of the vectors α_i. If all of the coefficients in this sum are positive (resp., negative), we call α a *positive root* (resp. *negative root*). We will denote by Π the set of positive roots (so that, clearly, $-\Pi$ comprises the set of negative roots). Although we will not prove it, Φ is the disjoint union of Π and $-\Pi$.

Given a collection T of roots in some root system Φ, we may consider the subgroup W_T of W generated by the reflections r_α corresponding to $\alpha \in T$. This subgroup itself has a root system (sometimes called a *subsystem* of Φ), which we may denote by Φ_T:

$$\Phi_T = \{\alpha \in \Phi \mid r_\alpha \in W_T\}.$$

1.2.3 *Examples*

We now examine some of the simplest (and most useful) Coxeter groups from both points of view introduced above.

In case W is finite, we can realize W as an orthogonal linear reflection group in some finite dimensional Euclidean space (see [Benson and Grove (1996)], [Bourbaki (1981)], or [Humphreys (1990)] for more details). There are a number of cases of interest.

The simplest examples of Coxeter groups are provided by the dihedral group D_n of order $2n$ and the symmetric group S_n (the group of permutations of $\{1, ..., n\}$) of order $n!$. Indeed, it is easy to see that D_n has presentation $\langle a, b \mid a^2, b^2, (ab)^n \rangle$ and that S_n has presentation

$\langle s_1, ..., s_{n-1} \mid (s_i s_j)^{m_{ij}} \rangle$, where $m_{ij} = 1 \Leftrightarrow i = j$, $m_{ij} = 2 \Leftrightarrow |j - i| \geq 2$, and $m_{ij} = 3 \Leftrightarrow |i - j| = 1$ (with arithmetic modulo $n - 1$). These groups furnish actions upon Euclidean space which are easily understood.

For instance, the group D_n corresponds to simple roots α and β in $\mathbb{E}^2$ which meet at an angle of π/n. As α and β we may choose unit vectors with a common vertex at the origin; the product $ab \in W$ then corresponds to a rotation about the origin through an angle of $2\pi/n$. (We have recaptured the realization of D_n as the group of symmetries of a regular n-gon.)

The action of S_n (as presented above) on $\mathbb{E}^n$ is nearly as simple. Consider the standard orthonormal basis $\{e_1, ..., e_n\}$ of $\mathbb{E}^n$. We let r_i act by permuting the ith and $(i + 1)$st elements of this basis ($r_i(e_i) = e_{i+1}$ and $r_i(e_{i+1}) = e_i$), leaving the remaining basis vectors fixed. It is easy to show that α_i is a unit vector pointing in the direction $e_i - e_{i+1}$, and that the line spanned by the vector $e_1 + \cdots + e_n$ is fixed (pointwise) by every r_i, and therefore by every element of W. Therefore S_n also acts on the hyperplane orthogonal to the vector $e_1 + \cdots + e_n$, and the action on this subspace of $\mathbb{E}^n$ has no nonzero fixed points (such an action is called *essential*).

Remark 1.3   In the literature on Coxeter groups and elsewhere, the dihedral group D_n is often denoted by $I_2(n)$, and the symmetric group S_n by A_{n-1}, in accordance with the notation commonly used in Lie theory.

There are a number of other finite Coxeter groups, comprising two other infinite classes of groups, as well as six "exceptional" groups. Some of these arise as the groups of symmetry of polyhedra of appropriate dimensions (see Exercise 4), and others can be realized in more exotic fashions. The reader may consult [Benson and Grove (1996)] or Chapter 2 of [Humphreys (1990)] for more details.

Before leaving the finite Coxeter groups behind, we mention that as simple as dihedral groups are, they will prove to be extremely important throughout our investigation. One reason for this is that many dihedral groups admit more than one Coxeter system. Indeed, if $n = 2k$ for k odd,

$$\langle c, d, g \mid c^2, d^2, g^2, (cg)^2, (dg)^2, (cd)^k \rangle \tag{1.4}$$

gives an alternate presentation for D_n. (The reader is asked to interpret this presentation geometrically as well.)

The simplest infinite Coxeter group is the infinite dihedral group, D_∞, with presentation

$$\langle a, b \mid a^2, b^2 \rangle. \tag{1.5}$$

Because the Coxeter matrix

$$\begin{bmatrix} 1 & -1 \\ -1 & 1 \end{bmatrix} \tag{1.6}$$

associated with this presentation is not positive definite, it cannot correspond to an inner product, and therefore we cannot put things in a nice Euclidean setting, as in the finite case. However, it is easy to find a natural "affine" action of D_∞ which generalizes the action of D_n by reflections in linear hyperplanes.

The real line, subdivided into intervals at integral points, is the "limit" of the sequence of regular n-gons, as n goes to infinity. we may embed this line into the Euclidean plane by identifying it with the line $\{(x, 1) \mid x \in \mathbb{R}\}$. For each $n \in \mathbb{Z}$, the point $(n, 1)$ determines a ray with the origin as its endpoint. Let X denote the union of all the triangular sectors determined by such rays, along with the rays themselves. (That is, X is the open upper half plane, along with the origin.) The action of a on X fixes pointwise the ray through $(0, 1)$ and exchanges sectors (rays) which are "mirror images" with respect to this ray. The action of b on X fixes pointwise the ray through $(1, 1)$ and exchanges sectors (rays) which are "mirror images" with respect to this ray.

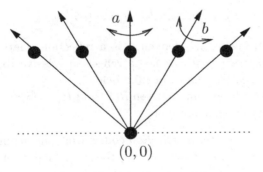

$(0, 0)$

Fig. 1.2 X, on which D_∞ acts by "linear reflections"

Remark 1.4   The space X described in the previous paragraph is known as the *Tits cone*, and is described more carefully in 1.5.2.

If we do not demand that W act by "linear" reflections, but only by affine reflections, we need not embed the real line into $\mathbb{E}^2$. Indeed, D_∞ already acts on the real line by affine reflections. In this way the group

D_∞ can be viewed as a two-dimensional *affine Euclidean reflection group*; it is the only such group in two dimensions. (There are three such groups in three dimensions, corresponding to tessellations of the Euclidean plane by regular hexagons; squares and octagons; and squares, hexagons, and dodecagons. A complete list of all affine Eucliean reflection groups, corresponding to positive semidefinite symmetric bilinear forms, is given in Chapter 4 of [Humphreys (1990)].)

1.3   Basic properties of Coxeter groups

1.3.1   *Reflections*

We return now to the combinatorial definition of the Coxeter group.

Let (W, S) be a Coxeter system as defined in 1.2.1. Any element of the form wsw^{-1} for $s \in S$ and $w \in W$ is known as a *reflection* of the system (W, S). (The elements of the fundamental generating set S are often called *simple reflections*.) The set of all reflections of the system will be denoted by $R(S)$. The set $R(S)$ does indeed depend on S.

For instance, consider the presentations of D_n ($n = 2k$, k odd) given in 1.2.3. An automorphism ϕ of D_n may be defined by

$$\phi(c) = a, \phi(d) = ababa, \phi(g) = (ab)^k. \tag{1.7}$$

In the second system given, the element g is a reflection relative to the generating set $\{c, d, g\}$, whereas g is not a reflection relative to the first system (which is most easily seen from 1.3.2 below).

We will explore the relationship between $R(S)$ and $R(S')$ for two systems (W, S) and (W, S') throughout the text.

Remark 1.5   Clearly, the term *reflection* comes from the geometric notion of the same name. Indeed, performing the construction of W indicated in 1.2.2, we see that the elements r_i of W act by reflection across the codimension-1 subspace H_i. Moreover, if w is any element of W defined as in 1.2.2, it is easy to show that wr_iw^{-1} acts by reflection across the codimension-1 subspace orthogonal to the vector $w\alpha_i$.

1.3.2   *Lengths of words and geodesics*

Let (W, S) be a Coxeter system, and let $w \in W$. It may be possible to express w as a product of simple reflections $s_1 \cdots s_n$ in more than one fashion. Let n be the smallest number for which the equality $w = s_1 \cdots s_n$ holds in the group W. The *length of w with respect to S*, denoted $l_S(w)$, is defined to be n. (When the generating set S is clear, we may write merely $l(w)$ for the length of w.) If $w =_W s_1 \cdots s_n$ and $l_S(w) = n$, we say the product $s_1 \cdots s_n$ is *reduced*, or that it is a *geodesic* expression for w.

The following proposition sums up some of the elementary properties of the length function. The verification of these facts is simple, and is left as an exercise.

Proposition 1.1 *Let $l(w)$ denote the length of $w \in W$ with respect to the generating set S, and let $w, w' \in W$.*

1. $l(w) = l(w^{-1})$.
2. $l(w) = 0 \Leftrightarrow w = 1$.
3. $l(ww') \leq l(w) + l(w')$.
4. $l(ww') \geq l(w) - l(w')$.
5. $l(w) - 1 \leq l(ws) \leq l(w) + 1$ *for every $s \in S$.*

There is very nice interplay between the length function $l(w)$ and the action of W upon the vector space V (and its corresponding root system Φ) as defined in 1.2.2. Following [Humphreys (1990)], we denote by $n(w)$ the number of positive roots in Φ sent to negative roots of Φ by the action of w; that is, $n(w) = |(\Pi \cap w^{-1}(-\Pi))|$. We refer the interested reader to Chapter 5 of [Humphreys (1990)] for a proof of the following theorem.

Theorem 1.1 *Let (W, S) be a Coxeter group, and let $w \in W$. Then $l(w) = n(w)$.*

The following is an important special case ($w = s_i$ for some $i \in I$) of the theorem.

Proposition 1.2 *Suppose that α_i is the simple root in Φ corresponding to $s_i \in S$. Then the set $\Pi \setminus \{\alpha_i\}$ is stable under the action of s_i. (That is, $s_i(\Pi \setminus \{\alpha_i\}) = \Pi \setminus \{\alpha_i\}$.)*

Proof. Let λ be a positive root, $\lambda \neq \alpha_i$. Then λ is expressible as a linear combination $\sum_{j \in I} a_j \alpha_j$, where for some $j \neq i$, $a_j > 0$. Applying the reflection s_i to both sides of the equation $\lambda = \sum_{j \in I} a_j \alpha_j$ shows that $s_i(\lambda) = \lambda - 2(\lambda, \alpha_i)\alpha_i$ is a linear combination involving the same coefficients for all α_j, $j \neq i$. Thus (since every root is either negative or positive) $s_i(\lambda)$ is again positive. Now $s_i(\lambda) \neq \alpha_i$, for then $\lambda = s_i s_i(\lambda) = s_i(\alpha_j) = -\alpha_j$, contradicting positivity. Therefore $\lambda \in \Pi \setminus \{\alpha_i\}$, and s_i acts by permuting the elements of $\Pi \setminus \{\alpha_i\}$. $\square$

From Proposition 1.2 we obtain the following useful result:

Proposition 1.3   *Suppose that α_i is the simple root in Φ corresponding to $s_i \in S$, and that $w \in W$. Then the following are true regarding the function $n(w)$.*

1. $w(\alpha_i) > 0 \Rightarrow n(wr_i) = n(w) + 1.$
2. $w(\alpha_i) < 0 \Rightarrow n(wr_i) = n(w) - 1.$
3. $w^{-1}(\alpha_i) > 0 \Rightarrow n(r_i w) = n(w) + 1.$
4. $w^{-1}(\alpha_i) < 0 \Rightarrow n(r_i w) = n(w) + 1.$

Proof.   If $w(\alpha_i) > 0$, then ws_i negates each of the roots $s_i(\alpha)$, where α is a positive root negated by w. (That all such roots are positive follows from Proposition 1.2.) Also, $ws_i(\alpha_i) \in -\Pi$ and α_i is not negated by w. Thus $n(ws_i) = n(w) + 1$. Similar arguments apply for the remaining cases, as the reader is encouraged to check. $\square$

1.3.3   *Parabolic subgroups*

Let (W, S) be a Coxeter system. For any subset $T \subseteq S$, we define the subgroup $W_T \leq W$ by taking as a generating set T and by applying only those relators from R which involve only letters of T. This subgroup is called the *standard parabolic subgroup* generated by T. Any conjugate wW_Tw^{-1} $(w \in W)$ is called a *parabolic subgroup* of W. (When T consists of a single generator t, we often write simply W_t instead of $W_{\{t\}}$.) Whether or not a group is a parabolic subgroup of W clearly depends on the system (W, S) that we choose.

Parabolic subgroups behave nicely with respect to intersection, as the following result demonstrates.

Proposition 1.4   *Let $T_1, ..., T_k \subseteq S$, and let $T = \bigcap_{i=1}^{k} T_i$. Then*

$$W_T = \bigcap_{i=1}^{k} W_{T_i}.$$

Proposition 1.4 follows from the fact that any two geodesic words representing the same group element must contain the same letters of the generating set S (*cf.* Proposition 1.5).

If W_T is a finite group, we call W_T a *spherical subgroup* of W. We may call T spherical as well. Since it is clear (prove!) that the full subgraph of $\mathcal{V}$ induced by the vertices of T is complete, we often say that T is a spherical *simplex* of $\mathcal{V}$. (This justifies our frequent use of letters typically reserved for simplices, such as σ, τ, *etc.*, to denote these collections.) If W_T is spherical and no spherical $T' \subseteq S$ properly contains T, we say that T is a *maximal spherical simplex* .

Spherical subgroups (and in particular, the maximal ones) play an important role in the theory of Coxeter groups. It is important that we be able to recognize the spherical subgroups of a given group W. (The enumeration of all spherical Coxeter groups was first done by Coxeter himself, in [Coxeter (1935)].) Clearly, any vertex generates a spherical subgroup, as does any edge. To detect three-generated spherical subgroups, we will make frequent use of the following lemma, often without mention.

Lemma 1.1   *Let* (W, S) *be a Coxeter system with diagram* $\mathcal{V}$. *Let the distinct vertices* $T = \{s_1, s_2, s_3\}$ *be form a triangle in* $\mathcal{V}$, *and let* $\{k, l, m\}$ *be the multiset of edge labels for this triangle. Then* W_T *is spherical if and only if*

$$\{k, l, m\} \in \Big\{ \{2, 3, 4\}, \{2, 3, 5\}, \{2, 4, 5\} \Big\} \cup \Big\{ \{2, 2, n\} \mid n \in \{2, 3, ...\} \Big\}.$$

The following lemma (proven below in 1.5.4) will also be used frequently.

Lemma 1.2   *Let* (W, S) *be a Coxeter system, and let* G *be an arbitrary finite subgroup of* W. *Then there exists a spherical subgroup* $W_T \leq W$ *and a group element* $w \in W$ *such that* $G \leq wW_Tw^{-1}$.

In particular, the conjugates of maximal spherical subgroups of W are the maximal finite subgroups of W.

As an application of this lemma, let us prove the following fact, in which W is realized (as in 1.2.2) as a group of orthogonal transformations in the vector space V, with associated root system Φ:

Lemma 1.3   *Let* $T \subset \Phi$ *be a finite set of roots closed under the action of* W. *Then the subgroup of* W *generated by the reflections in the roots* $T' = \Phi \cap \operatorname{span}(T)$ *is a finite parabolic subgroup of* W.

This will be used in Chapter 7 when we examine the automorphism group of certain Coxeter groups.

Proof.   Because T is finite, $\operatorname{span}(T)$ is a Euclidean vector subspace of V, and T' is also finite. At this point, an application of Lemma 1.2 and an unraveling of the definitions gives the desired result.   □

Given a system (W, S), we denote by $\mathcal{S}(S)$ (or simply by $\mathcal{S}$ when no confusion will arise) the collection of subsets T of S for which W_T is a spherical subgroup with respect to the generating set S. $\mathcal{S}(S)$ is a partially ordered set (poset) under inclusion. We denote by $W\mathcal{S}(S)$ (or by $W\mathcal{S}$) the set of *spherical cosets* $\{W/W_T \mid T \in \mathcal{S}(S)\}$. This set too is a poset under inclusion (see Exercise 8). We will have use for these collections in Section 1.5.3.

A well-known result from the general theory of Coxeter groups asserts that (W_T, T) is itself a Coxeter system, with the presentation described above. This fact highlights a possible ambiguity concerning the length function. To be precise, if $w \in W_T$ for some T properly contained in S, it is conceivable that $l_S(w) < l_T(w)$. However, this turns out to not be the case:

Lemma 1.4 *Let W_T be a standard parabolic subgroup of W, $W_T \neq W$. Then for any $w \in W_T$, $l_T(w) = l_S(w)$.*

The proof of this lemma is left as an easy exercise. Because of this result, it makes no difference whether we measure length in the subgroup W_T or in the ambient group W.

Given a parabolic subgroup W_T of W, it is often desirable to determine a set of *minimal coset representatives* for W_T. Let $T \subseteq S$ and define W^T by

$$W^T = \{w \in W \mid l(ws) > l(w) \text{ for all } s \in T\}. \tag{1.8}$$

That is, W^T consists of those elements of W for which reduced expressions cannot end in any letter of T. W^T is in some sense "complementary" to W_T, as demonstrated by the following fact.

Lemma 1.5 *Suppose W_T is a parabolic subgroup and W^T is defined as above. Given $w \in W$ there are unique elements $u \in W^T$ and $v \in W_T$ such that $w = uv$. Moreover, $l(w) = l(u) + l(v)$ holds, and u is the unique element of shortest length in the coset wW_T.*

The proof of this lemma too is left as an exercise. One means of proving it requires the Deletion Condition, a fundamental fact concerning Coxeter groups to which we now turn.

1.3.4 *The Deletion Condition*

The *Deletion Condition* not only provides a great deal of information concerning the reduced expressions for a given element $w \in W$; it also serves to characterize Coxeter groups in a very well-defined sense (Theorem 1.3).

The Deletion Condition is logically equivalent to the proposition known as the *Exchange Condition*. We will not prove the equivalence of these conditions, directing the reader to [Bourbaki (1981)] or [Humphreys (1990)] for further information. However, a simple proof of the Deletion Condition will be possible at the end of the next section.

Theorem 1.2 *Let (W, S) be a Coxeter system, and let $w \in W$.*

1. **[Deletion Condition]** *Suppose that $w = s_1 \cdots s_n$ holds in W, and let $l(w) < n$. Then there exist indices i and j satisfying $1 \leq i < j \leq n$ such that $w = s_1 \cdots s_{i-1} s_{i+1} \cdots s_{j-1} s_{j+1} \cdots s_n$ holds in w.*
2. **[Exchange Condition]** *Suppose that $w = s_1 \cdots s_n$ holds in W, where this expression is not necessarily reduced. Suppose that $l(ws) < l(w)$ for some $s \in S$. Then for some index i, $ws = s_1 \cdots s_{i-1} s_{i+1} \cdots s_n$. Moreover, if the expression is reduced, this index is unique.*

We note without proof here the following fact, which underscores the importance of the Deletion Condition to the theory of Coxeter groups.

Theorem 1.3 *Let G be a group generated by a set S of involutions (that is, S consists of elements of order 2). Then (G, S) is a Coxeter system if and only if G satisfies the Deletion Condition with respect to the set S. (That is, if $g = s_1 \cdots s_n$ in G and $l(g) < n$, then g can also be expressed as a product obtained from $s_1 \cdots s_n$ by omission of exactly two letters s_i, s_j.)*

More general groups satisfying the Deletion Condition are considered in Exercise 17.

In order to prove the Deletion Condition, we introduce a tool which will be useful later as well.

1.4   Van Kampen diagrams

In this section we present a brief account of the first powerful method from combinatorial group theory which we will have occasion to use. After defining van Kampen diagrams and stating a few fundamentral results concerning them, we will use them immediately to give an easy proof of the Deletion Condition. (This proof is originally due to A. Yu. Ol'Shanksii.) We will make regular use of van Kampen diagrams in the following chapters. For further applications of these diagrams to the study of Coxeter groups and related groups, the reader is encouraged to read [Appell and Schupp

(1983)], [Kapovich and Schupp (preprint)], and [Schupp (preprint)]. For the general theory of van Kampen diagrams, consult [Lyndon and Schupp (1977)] or [Ol'Shanskii (1991)].

Suppose that we are given a group presentation $G = \langle S \mid R \rangle$ where R is *symmetrized*. By this we mean that R is closed under taking inverses and cyclic permutations, and all elements of R are cyclically reduced.

A *map* Δ is a finite, connected, planar graph. Unlike the underlying graph of a Coxeter diagram, we regard Δ as a *directed graph*; that is, each edge comes equipped with an orientation, and its endpoints are referred to as its *initial* and *terminal vertices*. We will further regard a map as a planar 2-complex, considering also the 2-dimensional faces whose boundaries are formed by the edges of the graph Δ. Note that there is one unbounded face; namely, the one whose boundary is the perimeter of the graph Δ itself.

We define a *van Kampen diagram* Δ over the symmetrized presentation $\langle S \mid R \rangle$ (often simply called an R-diagram) to be a map with the following properties:

1. If e is an edge in Δ, then e is labeled by an element $\mathrm{lab}(e) = s$ in $S^{\pm 1}$, and $\mathrm{lab}(e^{-1}) = s^{-1}$.
2. If D is a face in Δ with boundary given by the edges $e_1, e_2, ..., e_k$, then the word $\mathrm{lab}(e_1)\mathrm{lab}(e_2) \cdots \mathrm{lab}(e_k)$ is (letter-for-letter; that is, as a word in the free group on S) a word in R.

Figure 1.3 illustrates this definition.

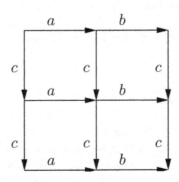

Fig. 1.3   A van Kampen diagram over the presentation $\langle a, b, c \mid aca^{-1}c^{-1}, bcb^{-1}c^{-1} \rangle$

Remark 1.6   In more general applications, $\mathrm{lab}(e)$ is merely assumed to be a word in the letters of $S^{\pm 1}$. Subdivision of edges of Δ (with concomitant subdivision of labeling words) then yields an R-diagram Δ as we have defined it.

We call the 2-dimensional faces of a van Kampen diagram *cells*. Much as products in the free group may often be reduced by cancellation of adjacent mutually inverse letters, a van Kampen diagram Δ may be put in a simpler form by cancellation of adjacent mutually inverse cells. That is, suppose that Δ contains two cells Π_1 and Π_2 for which at least one edge e appears in the boundary of both cells. Furthermore, suppose that the labels on the edges of Π_1, reading clockwise starting with $\text{lab}(e)$, give the same word in the free group on S as the labels on the edges of Π_2, reading counterclockwise starting with $\text{lab}(e)$. We may perform a *reduction* on Δ by removing the two cells Π_i, thus creating a "hole", and then "sewing up" the hole by identifying the edges on the boundaries of Π_1 and Π_2 which were not held in common (in the obvious fashion). This action does not effect the remaining cells of the diagram Δ, and we may perform many reductions, in sequence, on the same diagram. If no such reduction is possible, we say that Δ is *reduced*.

Combined with the method of small cancellation theory (see [Lyndon and Schupp (1977)]), van Kampen diagrams provide a powerful tool for proving theorems about groups with certain presentations. The most fundamental fact is the following theorem, often known as van Kampen's Lemma. For a proof, try Exercise 14, or see [Lyndon and Schupp (1977)].

Lemma 1.6 [**van Kampen's Lemma**] *Given a symmetrized group presentation* $G \cong \langle S|R \rangle$ *and a word w in the letters S, the element $\bar{w}$ represented by w is trivial in G if and only if there is a reduced R-diagram Δ whose boundary is labeled (letter-for-letter) by the word w.*

We now use van Kampen diagrams to analyze a Coxeter system (W, S). Since every generator $a \in S$ is its own inverse, we may consider an R-diagram over the symmetrized version of the presentation $\langle S \mid R \rangle$ as an undirected graph in which each edge is labeled by a single generator. For the remainder of this section, let Δ be such a van Kampen diagram.

Since each relator in the set R is of even length, we may speak unambiguously of the edge *opposite* a given edge on a face whose boundary label is given by a relator $r \in R$. Consider any edge on the boundary of Δ described above. This edge lies on the boundary of exactly one bounded face of Δ, and we may determine the edge that is opposite the original edge on this face. By planarity this edge either lies on the perimeter of Δ

or in exactly one other bounded face of Δ. If the latter condition holds, we may once more select the edge opposite this edge in this next bounded face. Continuing this process until we have reached another edge on the perimeter of Δ, we obtain the *band* beginning at the first edge chosen.

As a warm-up to the proof of the Deletion Condition, we prove the following fact:

Lemma 1.7   *Consider Δ as above, and construct a band as described above. Then the band does not cross itself.*

The forbidden situation is indicated in Figure 1.4.

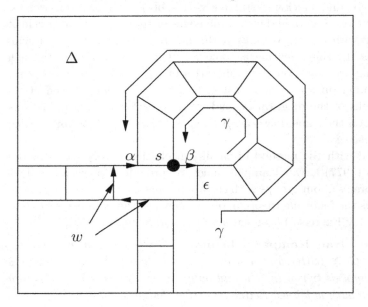

Fig. 1.4   A self-crossing band

Proof.   Suppose by way of contradiction that there is a band which self-crosses. We use the notation indicated in Figure 1.4. In particular, w is the word in the letters s and s' which makes up one half of the relator at which the band self-intersects, and therefore $\bar{w}\bar{\alpha}s\bar{\beta}\epsilon = 1$, where $\epsilon \in \{s, s'\}$ is the letter occurring at the end of the relator face farther from the boundary of the diagram Δ.

We claim that this presents a contradiction to the order of ss'. Indeed, note that the word γ can be read both on the "inside" of the arc formed by the band, as shown, and on the "outside". Therefore, reading from the basepoint indicated by a dot in Figure 1.4, we see that

$$\bar{\beta}\bar{\gamma} = s\bar{\alpha}^{-1}\bar{w}\bar{\gamma}s = 1.$$

Thus $\bar{\alpha}\bar{\beta} = \bar{w}$. But this is nonsense, as w is geodesic and $\alpha\beta$ is strictly shorter than w. $\qquad\square$

We now give the promised proof of the Deletion Condition.

Proof. Suppose that w is a freely reduced word representing $\bar{w}$ so that w is not geodesic. Let u be a geodesic word representing this same group element, and consider a van Kampen diagram Δ, as shown in Figure 1.5, with w labeling the "top", and u labeling the "bottom" of Δ. Such a diagram exists, because of van Kampen's Lemma. (We assume here that both wu^{-1} and $u^{-1}w$ are freely reduced. There are further subtleties regarding the exact appearance of Δ; these are left to the reader to resolve in Exercise 10.)

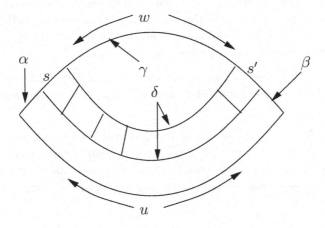

Fig. 1.5 The proof of the Deletion Condition

For any edge lying in the "top", labeled by w, we can construct a band as described above, which does not intersect itself (by Lemma 1.7) and which terminates somewhere on the boundary of Δ. We claim that for some edge lying in w, the corresponding band must terminate at another edge lying in w. Indeed, because u is strictly shorter than w, it cannot be that every band beginning in w terminates inside u.

Therefore, as shown in Figure 1.5, there is some band which begins and ends at different edges in w. Adopting the notation used in that figure, $\bar{\gamma} = \bar{\delta}$, and therefore

$$\bar{w} = \bar{\alpha} s \bar{\gamma} s' \bar{\beta} = \bar{\alpha} s \bar{\delta} s' \bar{\beta} = \bar{\alpha} \bar{\delta} \bar{\beta} = \bar{\alpha} \bar{\gamma} \bar{\beta},$$

and $\alpha\gamma\beta$ is obtained from w by the removal of exactly two letters (s and s'). $\qquad\square$

As another application of van Kampen diagrams, let us prove the following fact.

Proposition 1.5 *Let (W, S) be an even Coxeter system, and let $w \in W$. If the length of w is k and $s_1 s_2 \cdots s_k$ and $s_1' s_2' \cdots s_k'$ are both words in the letters of S which represent w, then every letter $s \in S$ appears the same number of times in the first word as in the second.*

Proof. Consider a van Kampen diagram Δ whose "top" edge is labeled $s_1 \cdots s_k$ and whose "bottom" edge is labeled $s_1' \cdots s_k'$. Let $s_i = s$. Because (W, S) is even, the band beginning at s_i must end at another occurrence of s. If this band ends at some edge in the top of Δ, we obtain a contradiction to the length of w. Thus the band ends at an occurrence of s in $s_1' \cdots s_k'$. Since s_i was arbitrary, the proof is complete. $\qquad\square$

Proposition 1.5 is a special case of the more general fact that any two geodesic words for the same group element must contain the same generators.

Remark 1.7 A large number of the fundamental results contained in [Bourbaki (1981)] and [Humphreys (1990)] are very easily proven using van Kampen diagrams. Further application of these diagrams will be found in this chapter's exercises, and throughout the remainder of the text.

1.5 Other geometric viewpoints: the Coxeter complex and the Davis complex

We conclude this chapter by considering a pair of geometric objects associated to a given Coxeter system (W, S), on which W acts in a natural fashion. (A third object, known as a *chamber system*, will be introduced in Chapter 6.)

The first construction is more "classical". The objects considered are called *Coxeter complexes*. The study of Coxeter complexes leads to the theory of buildings, developed primarily by J. Tits and his colleagues in the 1960s and 1970s, and detailed for instance, in [Brown (1989)] and [Ronan (1989)]. (Buildings have proven useful in algebraic geometry, CAT(0) geometry, and combinatorics, among other fields.)

The second construction is more recent, and was developed by M. Davis ([Davis (1983)]), building upon the work of E. B. Vinberg in [Vinberg (1971)]. (The more general spaces defined in [Vinberg (1971)] will be examined in 4.3.4.) The result of this construction will be called the *Davis complex* corresponding to a given system (W, S).

Before describing either of these objects, we provide a brief review of (abstract) simplicial complexes.

1.5.1 *Simplicial complexes*

Recall that an *abstract simplical complex* C with *vertex set* V is a collection Σ of finite subsets of C (each of which is called a *simplex* in C) satisfying the following conditions:

1. every singleton set $\{v\}$ ($v \in V$) lies in Σ, and

2. if $\sigma \in \Sigma$ and $\tau \subseteq \sigma$, then $\tau \in \Sigma$. (In this case, τ is called a *face* of σ.)

The *rank* of a simplex σ is defined to be its cardinality as a set; the *dimension* of $\sigma \in \Sigma$ is defined to be one less than its rank. A complex C is called *finite dimensional* if there is a finite upper bound on the dimension of its simplices. Clearly any simplicial complex Σ is a poset, where order is defined by the face relation: $\tau \leq \sigma \Leftrightarrow \tau \subseteq \sigma$.

The *geometric realization* $|C|$ of a an abstract simplicial complex C is constructed by taking the union over all simplices $\sigma \in \Sigma$ of the *open geometric simplices* $|\sigma|$, where $|\sigma|$ is the interior of a standard Euclidean simplex of the same dimension as σ. More precisely, we may begin with a vector space with basis V. We then have (for each $\sigma \in \Sigma$) an open simplex $|\sigma|$ consisting of the convex combinations

$$\sum_{x \in \sigma} \lambda_x x,$$

where $\sum_{x \in \sigma} \lambda_x = 1$ and $\lambda_x > 0$ for all $x \in \sigma$. C (under the face relation) is isomorphic to the poset of *geometric* simplices of the complex $|C|$ so constructed (under set inclusion).

If the data we are given to begin with describe a poset rather than an abstract simplicial complex, we may define a related simplicial complex in the following fashion. Given the poset P, let $\mathrm{Ord}(P)$ be the collection of all finite chains (totally ordered subsets) of P. Considering each such chain as a simplex, $\mathrm{Ord}(P)$ is an abstract simplicial complex, isomorphic to the poset of simplices of its geometric realization, which we will denote by $\mathrm{Geom}(P)$. Note that, for instance, each element of P corresponds to a vertex of $\mathrm{Geom}(P)$, and a set of vertices of $\mathrm{Geom}(P)$ spans a simplex if and only if the corresponding elements of P form a totally ordered subset of P.

Given any lower set $\downarrow_P x = \{y \in P \mid y \leq x\}$ or upper set $\uparrow_P x = \{y \in P \mid y \geq x\}$ in P, we may define subcomplexes of $\mathrm{Geom}(P)$ by $\mathrm{Geom}(\downarrow_P x)$ and $\mathrm{Geom}(\uparrow_P x)$, which are called, respectively, *faces* and *dual faces* of $\mathrm{Geom}(P)$.

Recall that an action of a group G on a simplicial complex C is called *simplicial* if for every $g \in G$ and for every simplex $\sigma \in C$, $g \cdot \sigma$ is again a simplex of C.

1.5.2   *The Coxeter complex*

Given (W, S), we define a *special coset* to be any coset wW_T, where $w \in W$ and $T \subseteq S$. We define a poset P whose elements are the special cosets. For $w_1 W_{T_1}, w_2 W_{T_2} \in P$, we say that $w_1 W_{T_1} \leq w_2 W_{T_2}$ if and only if $w_2 W_{T_2} \subseteq w_1 W_{T_1}$. The Coxeter complex $X = X(W, S)$ is now defined to be the geometric realization $\mathrm{Geom}(P)$ of this poset.

X possesses a great deal of structure beyond its simpliciality. The maximal dimensional simplices of X are called its *chambers*, and the codimension-1 faces of a chamber are called its *mirrors*. The support of a given mirror (that is, the affine subspace determined by the mirror) is called its *wall*. The walls of X are thus codimension-1 (affine) hyperplanes in the space in which X is realized. The complement of each wall consists of two disjoint open sets, each called a *half-space* of X. It can be shown that every mirror is a face of exactly 2 chambers, and that any two chambers can be connected by a sequence of "adjacent" chambers known as a *gallery*. These properties make X into what is known as a *chamber complex*.

The group W acts on X in a natural way: $w \cdot w_1 W_T = (ww_1)W_T$. This action turns out to be simply transitive on the set of chambers (that is, the action is transitive and the only element w taking any given chamber to itself is the trivial element). We can identify some chamber C as the *fundamental chamber*, and all other chambers are translates of C under W's action. Each $s \in S$ then acts as a "reflection" in some wall H_s supporting a codimension-1 face of C. A single vertex of C is moved by this reflection, so we can unambiguously label each vertex of C with that generator. Moreover, each vertex v of X can be labeled by an element of S by translating v back to a (unique!) vertex of C and giving v the label of that vertex. This yields a canonical *labeling* of X which makes X into a *labeled* chamber complex, in which the given labeling respects the action of W on X in a very natural way.

For the precise definitions of the terms and concepts introduced above, the reader is encouraged to consult [Brown (1989)]. We will examine X in more detail as it becomes necessary to do so (in Chapter 7). We content ourselves now with a few examples.

If W is a finite Coxeter group, then W can be realized as a subgroup of the group $GL(V)$ of invertible linear transformations on some finite-dimensional Euclidean vector space V. In this case, it turns out that the simplices of $X(W, S)$ are in one-to-one correspondence with the simplices of V defined by the intersection of the hyperplanes in which the elements of W act by reflections. (That is, this cell structure on V is isomorphic, as an abstract simplicial complex, to X. If, for example, $W = D_n$ with the usual presentation, $V = \mathbb{E}^2$ is decomposed into $2n$ triangular sectors, each having an angle $\frac{\pi}{n}$ at the origin, as shown in Figure 1.6. *Cf.* 1.2.3.)

In case W is infinite, X is not quite so simple. Although in general V can still be represented as a subgroup of the linear group $GL(V)$ for some finite-dimensional vector space V, V itself may no longer be the carrier for the complex X. However, we may identify X with a certain subset of the dual space V^*, properly subdivided into simplices. This subspace, known as the *Tits cone*, is defined most easily in terms of a fundamental domain for W's action.

Precisely, let C be defined by

$$C = \{f \in V^* \mid f(v) > 0 \text{ for all } v \in V\}. \tag{1.9}$$

(It is known that this set is not empty.) Then, given $w \in W$ and $f \in V^*$, we define $w \cdot f$ by $(w \cdot f)(v) = f(w^{-1}v)$ for all $v \in V$. Then the Tits cone is defined to be the union

$$\bigcup_{w \in W} w\bar{C}$$

where $\bar{C}$ is the closure of C in V^*.

As an example, consider the case of D_∞, examined in 1.2.3. Here C is the sector determined by the rays through the points $(0, 1)$ and $(1, 1)$,

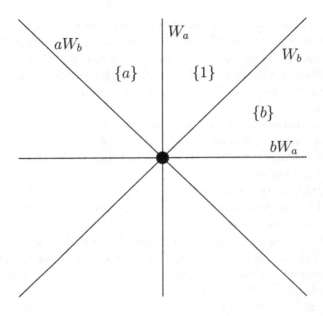

Fig. 1.6   The Coxeter complex $X(D_4, \{a, b\})$

and the Tits cone is the open upper half-space of $\mathbb{E}^2$, plus the origin. (The origin is the vertex of X corresponding to the group $W = D_\infty$ itself.)

More information on the Tits cone can be found in Section 5.13 of [Humphreys (1990)] and in [Howlett, *et al.* (1997)], for instance. It will play a major role in our work only in Section 7.2.

1.5.3   *The Davis complex*

We now define a different simplicial complex $\Sigma = \Sigma(W, S)$ on which a given Coxeter group W acts in a manner similar to the action on $X(W, S)$ above. The construction of Σ is due to M. Davis (in [Davis (1983)]), following a more general construction of E. B. Vinberg ([Vinberg (1971)]) which will be discussed briefly in Section 4.3, where it will be required in order to complete the proof of Theorem 4.4.

The complex Σ has the advantage that it can be given a metric that yields a very desirable geometry.

Theorem 1.4   *Let (W, S) be a Coxeter system. Then there is a locally finite simplicial complex $\Sigma = \Sigma(W, S)$ on which W acts simplicially, cocom-*

pactly, and properly by isometries. Moreover, Σ admits a natural piecewise Euclidean CAT(0) *metric.*

Recall that a group G is said to act *cocompactly* on the metric space X if there exists some compact subset K of X such that $G \cdot K = X$. The action is said to be *proper* if for every point $x \in X$, there is a number $r > 0$ such that the set

$$\{g \in G \mid g \cdot B(x, r) \cap B(x, r) \neq \emptyset\}$$

is finite. ($B(x, r)$ is the open ball of radius r around x.) Note that the action of W on the Coxeter complex $X(W, S)$ is proper, although not cocompact.

Recall the definitions of the posets $\mathcal{S} = \mathcal{S}(S)$ and $W\mathcal{S} = W\mathcal{S}(S)$ given in 1.3.3. Let $\Sigma = \text{Geom}(W\mathcal{S})$, $K = \text{Geom}(\mathcal{S})$, and $K_T = \text{Geom}(\uparrow_{\mathcal{S}} T)$. There is a natural inclusion of K into Σ induced by $T \mapsto 1 \cdot W_T$. The image of K in Σ under this identification is called the *fundamental chamber* of Σ. Since W acts in a natural way on $W\mathcal{S}$ (by $w \cdot w'W_T = (ww')W_T$, just as in the case of the Coxeter complex), we can extend this to an action of W on Σ. As in the Coxeter complex, the translates of K under this action are called the *chambers* of the action; the sets $K_s = K_{\{s\}}$ are called the *mirrors* of K. It is across the hyperplane in $\Sigma(W, S)$ which supports the mirror K_s that s acts as a reflection. Any translate of such a hyperplane will be called a *wall* of the complex $\Sigma(W, S)$. Any wall of $\Sigma(W, S)$ is exactly the fixed point set of some reflection in (W, S).

We call $\Sigma = \Sigma(W, S)$ the *Davis complex* associated to (W, S).

Remark 1.8 There are two fundamental differences between the Davis complex and the Coxeter complex. First, the Coxeter complex contains a vertex corresponding to *any* coset of a standard parabolic subgroup, whereas the vertices of the Davis complex correspond only to cosets of spherical subgroups. Second, the order in the poset of all special cosets is defined by reverse inclusion, which order is dual to that used in the definition of the Davis complex.

In case W is finite, it is often easy to visualize K and Σ, and this construction yields unsurprising results. For instance, suppose that W is the dihedral group D_n of order $2n$ with the usual presentation: $\langle a, b \mid a^2, b^2, (ab)^n \rangle$. Then Σ can be identified topologically with a $2n$-gon, along with its interior. Combinatorially, we subdivide the $2n$-gon into $4n$ 2-simplices, where K consists of the union of two of these simplices, and of which $2n$ translates fill out Σ. The barycenter of the subdivided $2n$-gon corresponds to the group W itself.

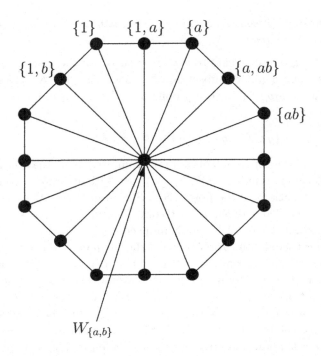

Fig. 1.7   The Davis complex $\Sigma(D_4, \{a, b\})$

The reader is invited to construct the Davis complex for a few other simple groups in the exercises.

The complex Σ has a number of very nice properties, some of which we will investigate in the course of our studies.

For instance, given a simplex σ in Σ, the subgroup $\mathrm{Stab}(\sigma)$ of W of elements which pointwise fix σ is a conjugate of a spherical subgroup of W. Indeed, a given simplex corresponds to a chain of spherical cosets $w_1 W_{T_1} < w_2 W_{T_2} < \cdots < w_n W_{T_n}$. It is easy to show that $\mathrm{Stab}(\sigma)$ is given by $w_1 W_{T_1} w_1^{-1}$. Using this and the fact that Σ carries a CAT(0) metric as defined in the next subsection, we will be able to prove Lemma 1.2, one of the most important intermediate results in this text.

We will now discuss the construction of the CAT(0) metric in very general terms, referring the reader to [Davis and Moussong (1999)] for a detailed study.

1.5.4 *The metric on* Σ

We first provide the definition of the CAT(0) condition, which requires a number of additional geometric notions.

For any real number κ there is (up to isometry) a unique complete, simply connected, Riemannian manifold M_κ^n of dimension n and constant sectional curvature κ. For $\kappa = 0$, M_κ^n is simply n-dimensional Euclidean space. For $\kappa > 0$, M_κ^n is spherical, and for $\kappa < 0$ (the most interesting case for many group theoretic purposes), M_κ^n is hyperbolic. These spaces are frequently called *model spaces* of a given curvature κ because they provide models for the behavior of spaces which exhibit similar curvature, either locally or globally.

Now suppose that X is a metric space with metric d. Let x, y be two points in X, and let α be a continuous function from the interval $[0, l]$ to X which satisfies $\alpha(0) = x$, $\alpha(l) = y$, and $d(\alpha(t), \alpha(t')) = |t - t'|$ for every $t, t' \in [0, l]$. In this case we call α a *geodesic* from x to y, and the image $\alpha([0, l])$ is called a *geodesic segment* from x to y, and may be denoted by $[x, y]$. These definitions makes precise the notion of the "shortest path" between two points in the space X.

Fix a nonpositive real number κ. (The following definitions can be modified to deal with the case in which X is positively curved, but since we are most interested in the case when $\kappa \leq 0$, we will not address the case in which $\kappa > 0$.)

Consider any three points x, y, z in X and construct a triangle Δ whose sides consist of geodesic segments between the respective vertices x, y, z. We can find a (geodesic) triangle Δ' in the model space $M = M_\kappa^2$, with vertices x', y', z', such that $d(x, y) = d(x', y')$, $d(y, z) = d(y', z')$, and $d(z, x) = d(z', x')$ (with obvious abuse of notation). Let a and b be any points on the triangle Δ, and let a' and b' be the corresponding points in Δ'. That is, for instance, if a lies on the geodesic segment $[x, y]$ and $d(a, x) = r$, then a' is the point lying on the geodesic segment $[x', y']$ such that $d(x', a') = r$. If for any choice of a and b

$$d_X(a, b) \leq d_M(a', b') \tag{1.10}$$

holds, we say that Δ satisfies the CAT(κ) inequality. If $\kappa \leq 0$ and (1.10) is satisfied for *every* geodesic triangle in X, X is said to be CAT(κ).

Remark 1.9 The term CAT(κ) was introduced by M. Gromov, and comes from the names E. **C**artan, A.D. **A**lexandrov, and V.A. **T**oponogov, three mathematicians who considered problems very closely related to the

CAT(κ) condition.

How does one impose a CAT(0) metric upon the Davis complex Σ? Consider a Coxeter system (W, S). Roughly speaking, we define the "shape" of the portions of Σ corresponding to spherical subgroups W_T in such a manner that when these portions intersect nontrivially, the shapes can be glued together isometrically. The geometry of the entirety of Σ is then induced by the fact that K is a fundmental domain for the action of W on Σ. The resulting metric is piecewise Euclidean, and the action of W on Σ is by isometries.

More precisely, suppose first that we are given a system (W, S) such that W is finite, and select any point x in the fundamental domain for the action of W as a reflection group in Euclidean space $\mathbb{E}^n$ (x may be given, for example, by specifying the distance from x to each of the hyperplanes corresponding to the generators of S). We define the *Coxeter cell* C to be the convex hull in $\mathbb{E}^n$ of the orbit Wx of x under the action of W. (For different choices of x, the exact shape of C will vary.) For example, suppose that $W = D_n$ is presented as $\langle a, b \mid a^2, b^2, (ab)^n \rangle$ and x is chosen to be equidistant the two hyperplanes which bound the fundamental domain consisting of a sector of $\mathbb{E}^2$ with angle π/n. Then C will be a regular n-gon.

The following lemma is straightforward (see [Davis and Moussong (1999)]).

Lemma 1.8 *If W is finite and C is a Coxeter cell corresponding to W, then the poset WS is isomorphic to the poset* Faces(C) *of faces of C by the correspondence $w \mapsto wx$.*

For example, a Coxeter cell corresponding to the dihedral group D_n is a $2n$-gon, each of whose edges have one of two lengths (in general), which alternate.

Now consider (W, S) for an arbitrary Coxeter group W. We fix once and for all a map f from S to the positive real numbers which will be used to determine the shape of the Coxeter cells C introduced below. (As above, we can imagine that this map prescribes a point x lying a predetermined distance from each generator's hyperplane.) For each spherical subgroup W_T of W, we define a piecewise Euclidean metric structure on Σ by identifying the face (as defined in 1.5.1) Geom($\downarrow_{WS} W_T$) with the Coxeter cell $C(W_T)$ corresponding to W_T which is defined by restricting the map f to T. Any translate of a face of Geom(W_T) also has shape determined by $C(W_T)$.

For example, consider the group

$$W = \langle a, b, c \mid a^2, b^2, c^2, (ab)^3, (bc)^3, (ac)^3 \rangle,$$

which we will hereafter refer to as $(3, 3, 3)$. The Davis complex for W is (combinatorially) a triangulation of the Euclidean plane by triangles. The Coxeter cells (i.e., the faces) corresponding to the 2-generated subgroups of W are hexagons, each of which falls into one of at most three isometry classes.

The action of W upon Σ is then clearly by isometries. It is, however, a non-trivial matter to show that the metric which is induced by this structure is CAT(0). The interested reader may consult [Davis and Moussong (1999)] for details.

Remark 1.10 In [Moussong (1996)], Moussong also characterizes word hyperbolic Coxeter groups. A group G is said to be *word hyperbolic* with respect to a particular generating set S provided the Cayley graph $\Gamma(G, S)$ is a hyperbolic metric space (relative to the usual path metric). This means, roughly, that $\Gamma(G, S)$ satisfies a certain negative-curvature condition. Although we know it to be true for Coxeter groups, it is an open question whether every word hyperbolic group acts properly and cocompactly by isometries on some CAT(0) metric space.

Because W acts so nicely on a CAT(0) space, W enjoys a number of desirable properties. For instance, the Conjugacy Problem for Coxeter groups is solvable (see Section 2.4). For now we content ourselves with a proof of Lemma 1.2.

Proof. We apply the following result (see [Bridson and Haefliger 1999], II.2) to the action of $G \leq W$ on the Davis complex $\Sigma(W, S)$:

Proposition 1.6 *Let G be a finite group of isometries of a complete CAT(0) space X. Then the set X^G of points of X fixed by G is a non-empty convex subspace of X.*

Our finite $G \leq W$ is a finite group of isometries of the CAT(0) space Σ, and the above proposition shows that its set of fixed points is non-empty. Given a point v in Σ fixed by G, G fixes the smallest simplex containing v. By the construction of Σ, the isotropy group of a simplex is a finite parabolic subgroup $wW_T w^{-1}$ (for some $T \subseteq S$). Thus $G \leq wW_T w^{-1}$ for this T. □

We will apply Lemma 1.2 almost immediately as soon as we begin to prove theorems concerning rigidity.

1.6   Exercises

1. Let W be a finitely generated Coxeter group. Prove that the set Φ of roots associated to a Coxeter group W as in 1.2.2 is finite if and only if W is finite.

2. Define Φ and Π as in 1.2.2. Prove that Π and $-\Pi$ exhaust Φ in case W is finite.

3. Verify all of the statements regarding the mapping r_i made in the paragraph containing (1.3).

4. Let $\{e_1, ..., e_n\}$ be the standard orthonormal basis of $\mathbb{E}^n$. The group W of order $2^n n!$ generated by exchanges $e_i \leftrightarrow e_j$ $(1 \leq i < j \leq n)$ and by exchanges $e_i \leftrightarrow -e_i$ $(1 \leq i \leq n)$ is a Coxeter group. Find a generating set of n reflections for this group, and compute the corresponding Coxeter matrix and diagram. (These groups, typically denoted B_n, are known as the *hyperoctahedral groups*, as they are the groups of symmetry of the regular octahedra of a given dimension.)

5. Prove Proposition 1.1.

6. Prove Lemma 1.4. (*Hint*: use the function $n(w)$ from 1.3.2.)

7. Prove that if (W, S) is a Coxeter system with finite group W, then there is a unique element Δ_W of greatest length in W. Moreover, show that Δ_W has order 2, and that for any $s \in S$, $\Delta_W s \Delta_W^{-1} = s'$ for some $s' \in S$. (The element Δ_W is called the *Garside element* or *Coxeter element* of W. The former term comes from the work of F.A. Garside, who produced pioneering work on certain Artin groups. Artin groups are discussed in Chapter 3 and elsewhere in the sequel.)

8. Prove that $WS(S)$ is a poset under inclusion, and show that $w_1 W_{T_1} \leq w_2 W_{T_2}$ if and only if $T_1 \subseteq T_2$ and $w_2^{-1} w_1 \in W_{T_2}$.

9. Prove Lemma 1.5. Also, state and prove the corresponding result for right cosets.

10. What subtleties are overlooked in our application of van Kampen diagrams to proving the Deletion Condition? Explain how to get around any such pesky details.

11. Describe (and draw, as well as you can!) the Davis complex for each of the groups Σ_n, $(\mathbb{Z}_2)^n$, and $(3, 3, 3)$, indicating in each case the fundamental

domain K and the walls corresponding to various reflections.

12. (For the definition of the *Cayley graph* of a group G relative to a generating set S, consult any text on group theory.) Let (W, S) be a Coxeter system. Prove that the Cayley graph embeds into the cell complex Σ' (defined by Coxeter cells as in 1.5.4) dual to the Davis complex $\Sigma(W, S)$. (This identification turns out to be a *quasi-isometric* embedding; that is, distances between points on the Cayley graph are distorted in a controllable fashion upon embedding into the Davis complex.)

13. Let (W, S) be a Coxeter system, with diagram $\mathcal{V}$. The *nerve* $N = N(W, S)$ of the system (W, S) is the simplicial complex which results from $\mathcal{V}$ by attaching a k-dimensional simplex according to the vertex set σ ($|\sigma| = k + 1$) whenever W_σ is spherical. Let $\Sigma(W, S)$ be given the cell structure determined by identifying each face of Σ with the Coxeter cell of the corresponding type (as in 1.5.4). For every vertex in Σ so cellulated, prove that $\mathrm{Lk}(v, \Sigma)$ is homeomorphic to N. (Recall that the *link* $\mathrm{Lk}(v, C)$ of a vertex v in the simplicial complex C is the simplicial complex consisting of a simplex σ' of dimension $k - 1$ for each k-simplex σ of C containing v in its closure.)

14. Prove van Kampen's Lemma (Lemma 1.6). (*Hint*: In one direction, you can induct on the number of cells in the given R-diagram. For the other implication, begin by expressing the word w as a product, in the free group on S, of conjugates of relators from R.)

15. Prove: if T is a spherical subset of S, the vertices of T induce a complete subgraph of $\mathcal{V}$.

16. Prove: for any two points x and y in a CAT(0) metric space, if a there exists a geodesic from x to y, then this geodesic is unique. (More is true, in fact, for it turns out that CAT(0) metric spaces are *geodesic*: there exists a geodesic between any two points. See [Bridson and Haefliger 1999] for details.)

17. A *right-angled Artin group* G is any group with generating set $A = \{a_i\}_{i \in I}$ and presentation

$$\langle A \mid R \rangle,$$

where $R = \{a_i a_j = a_j a_i\}$ for some (though not necessarily all) pairs $i \neq j$ in I. Note that every generator has infinite order, so that (G, A) is not a Coxeter system. Use van Kampen diagrams to show that G *does* satisfy the Deletion Condition with respect to A, so that Theorem 1.3 is not true if the generating set S given there does not consist of involutions. (The most general groups satisfying the Deletion Condition are known as weakly partially commutative Artin-Coxeter groups, and are investigated by S.A. Basarab in [Basarab (2002)].)

18.   Describe all subgroups of a given dihedral group, D_{2n}, noting any differences between the cases n odd and n even.

19.   Let W be realized as the group of orthogonal transformations generated by the reflections $\{r_i\}_{i \in I}$ in some vector space V. Let Φ be the associated root system. Prove that if $v \in V$ is a unit vector (with respect to the bilinear form defined in 1.2.2) such that $r_v \in W$, then $v \in \Phi$. (*Hint:* induct on the length of r_v and use results from 1.3.2.)

20.   With the same set-up as in the previous exercise, assume that Φ is irreducible. Prove that for any $v_1, v_2 \in \Phi$, there is $v \in \Phi$ such that $(v_i, v) \neq 0$, $i = 1, 2$.

21.   Let W, V, and Φ be as in the previous two exercises. In [Brink and Howlett (1993)], B. Brink and R. Howlett describe a partial order on the set $\Pi \subseteq \Phi$, the set of positive roots; this order was later extended (in [Howlett, *et al.* (1997)]) to all of Φ as follows. Let $\alpha, \beta \in \Phi$. We say that α *dominates* β, and write $\alpha \succeq \beta$, if

$$\{w \in W \mid w\alpha \in -\Pi\} \subseteq \{w \in W \mid w\beta \in -\Pi\};$$

that is, if every element of W negating α also negates β. The corresponding strict order is denoted by $\succ$. In [Brink and Howlett (1993)] it is shown that for all $\alpha, \beta \in \Phi$, $(\alpha, \beta) \geq 1 \Leftrightarrow \alpha \succeq \beta$ or $\beta \succeq \alpha$.

Suppose that $S = \{a, b, c\}$, with corresponding simple roots $\Delta = \{\alpha, \beta, \gamma\}$. Suppose moreover that each of m_{ab}, m_{ac}, and m_{bc} are at least 3. Show that there exist roots $\{\alpha_1, \beta_1, \gamma_1\} \subseteq \Phi$ such that $\alpha_1 \succ \alpha$, $\beta_1 \succ \beta$, and $\gamma_1 \succ \gamma$. (*Hint:* new roots as desired can be obtained from α, β, and γ under the action of the appropriate words of length 2.)

This result can be generalized to all 3-generated infinite irreducible Coxeter systems. Much more is known about the dominance order $\succeq$. For instance, it can be shown that the set of positive roots which are minimal with respect to the order (among all positive roots) is finite. In Chapter 7 we will say more about $\succeq$.

Chapter 2

Further properties of Coxeter groups

In this chapter we examine more closely a handful of topics which will be particularly useful later in our study.

We begin by providing a crash course on the basics of group (co)homology, emphasizing theorems which will be needed later. For instance, we highlight Poincaré duality, which we will apply in Chapter 4.

From there we turn to the computation of normalizers and centralizers of various subgroups of a given Coxeter group. Knowing the structure of these subgroups will help us both to prove rigidity for rather general classes of Coxeter groups (Chapter 6), and to understand the structure of $\text{Aut}(W)$ for certain groups W (Chapter 7).

Next, we examine the ways in which a given group W can be expressed in terms of simpler groups by means of free products with amalgamation. Such decompositions of W will prove useful in Chapter 6.

Finally, in Section 2.4, we show that Coxeter groups have both solvable Word Problem and solvable Conjugacy Problem. We do this not only to indicate the techniques involved, but also to underscore the difficulty of the Isomorphism Problem.

2.1 The (co)homology of a Coxeter group

We assume that the reader has basic knowledge of homology theory, and take as given the definitions of simplicial homology and and cohomology. (Knowledge of basic properties such as excision will also be assumed in Chapter 4.) A useful introduction to algebraic topology is [Hatcher (2002)]. We refer the reader to [Brown (1992)] for careful definitions and exposition of nearly all of the material introduced in this section.

2.1.1   *Group (co)homology*

In the 1930s Hurewicz proved that for topological spaces X of a certain type (namely, for aspherical spaces), the homotopy type of X is completely determined by its fundamental group. That is, if two aspherical spaces have the same fundamental group, π_1, then they have the same homotopy type, and therefore the same homology groups. In this case, one may as well forget about the spaces in question and consider only the group π_1. The corresponding homology groups comprise what deserve to be called the homology groups of the *group* π_1. Out of this insight was born the homological study of groups.

It soon became clear that a number of questions concerning groups which had always been considered in a strictly algebraic fashion could be approached using the new topological methods being developed by Hurewicz, Hopf, and Eilenberg and MacLane, among others. Today homological methods in group theory provide an almost indispensible tool for dealing with a number of seemingly group theoretic problems.

One of the first steps in applying such methods to a certain class of groups is *computing* the (co)homology groups, as they are defined algebraically. This can often be a daunting task, although there are "shortcuts". For example, the chore is often simplified if a $K(G, 1)$-complex for the group G can be found easily. Given G, a $K(G, 1)$-*complex* for G is any connected CW-complex X with a contractible universal cover such that $\pi_1(X) \cong G$. (For an overview of CW-complexes, see [Hatcher (2002)].) The usefulness of these complexes is demonstrated by the following result:

Theorem 2.1   *Suppose that X is a $K(G, 1)$-complex for the group G. Then the homology of X is identical to that of G.*

That is, the group's homology can be found by computing the homology of a space X naturally associated to it. (Given the manner in which group homology was born, this should come as no surprise!) Shortcuts of this sort are not always available to us, however. For instance, even when a $K(G, 1)$-complex can be found, its homology may be no more easy to compute than that of the group.

As an example, let (W, S) be a Coxeter system for the infinite Coxeter group W. Because W acts properly on $\Sigma = \Sigma(W, S)$ (see 1.5.3) and because for any finite subgroup $G \leq W$, the set of points of Σ fixed by G is contractible, Σ is called a *classifying space for proper W-actions*. (This means that Σ is universal in the sense that for any other CW-complex X

on which W acts properly, there is a more or less unique W-equivariant map from X to Σ, so Σ is as "small" as we have any hope of making it.) Now consider a torsion-free subgroup G of finite index in W. It can be shown that Σ is the universal cover of a $K(G, 1)$-complex. Very roughly, then, Σ has the same homology as the torsion free subgroup G. However, computing the homology of Σ is in general no easier than computing the homology of the group G itself.

In [Davis (1998)], M. Davis exploits the action of W upon the simplicial complex $\Sigma = \Sigma(W, S)$ in order to compute the (co)homology (with group ring coefficients) of an arbitrary Coxeter group W. The answer is given in terms of the relative (co)homology of the fundamental domain $K \subseteq \Sigma$ defined in 1.5.3:

Theorem 2.2 *Let $\mathbb{Z}W$ be the integral group ring of W. Then*

$$H^*(W; \mathbb{Z}W) \cong \bigoplus_{w \in W} H^*(K, K^{T(w)}). \tag{2.1}$$

Here,

$$K^T = \bigcup_{t \in T} K_t$$

for all $T \subseteq S$, and $T(w) = \{s \in S \mid l(ws) > l(w)\}$. (Thus $T(w)$ is the set of generators with which a geodesic representative for w cannot end.)

Theorem 2.2 actually holds if one replaces W by any subgroup of finite index throughout. Subgroups of finite index play a role in defining another statistic of a given group G, namely, its *virtual cohomological dimension*.

The *cohomological dimension* of a group G is the greatest number n such that $H^n(G; M)$ is not the trivial module, for some module M on which the group ring $\mathbb{Z}G$ acts. (If there is no such finite number, the cohomological dimension is defined to be ∞.) An important theorem due originally to J.-P. Serre states that every finite-index subgroup H of a given group G has the same cohomological dimension as G. Therefore we can unambiguously define the virtual cohomological dimension vcd(G) of G to be the cohomological dimension of any subgroup H of finite index. It can be shown that vcd(G) cannot exceed the dimension of any $K(G, 1)$-complex for G.

2.1.2   *Poincaré duality*

We will make use of one other homological property possessed by certain Coxeter groups. We open the discussion of *Poincaré duality* with a theorem:

Theorem 2.3   **[Poincaré duality]**   *Let M be a closed, orientable n-manifold. Then there is a natural isomorphism between the groups $H_c^k(M)$ and $H_{n-k}(M)$ for all k. (Here H_c^k is cohomology with compact support. That is, H_c^k is defined on cochains with finite support.)*

A group G is called a *Poincaré duality group* if G satisfies a homological duality condition similar to that highlighted in Theorem 2.3. For instance, if there is an integer n such that $H^i(G; \mathbb{Z}G) = 0$ for all $i \neq n$ and $H^n(G; \mathbb{Z}G)$ is infinite cyclic, then (provided G acts trivially on the latter module) we have the following identity:

$$H^i(G; M) \cong H_{n-i}(G; M),$$

for any module M. (*Cf.* Theorem 2.3.)

The above condition is satisfied in particular if G possesses a $K(G, 1)$-complex which is a closed, orientable manifold. In this case, therefore, G's homology groups can be readily computed from G's cohomology groups, and *vice versa*.

If G itself does not satisfy Poincaré duality (as defined above) but some finite-index subgroup $H \leq G$ does, we call G a *virtual Poincaré duality group*. It turns out that if any one subgroup of finite index satisfies Poincaré duality, then all other subgroups of finite index do so as well.

Poincaré duality is essential in proving Proposition 4.4, in which we compute the reflections of one system in terms of those of another. The Coxeter groups which satisfy virtual Poincaré duality turn out to be precisely those groups of type HM_n defined in 4.3.1.

2.2   Normalizers and centralizers

Let G be a group, and let H be a subgroup of G. The *centralizer* of H in G (which we will denote by $C_G(H)$, or simply $C(H)$ when the ambient group G is evident) is the set

$$\{g \in G \mid gh = hg \text{ for all } h \in H\}.$$

$C(H)$ is obviously a subgroup of G. The *normalizer* of H in G (denoted by

$N_G(H)$ or $N(H)$) is the set

$$\{g \in G \mid ghg^{-1} \in H \text{ for all } h \in H\}.$$

Obviously $N(H)$ too is a subgroup of G, $C(H) \leq N(H)$, and $N(H)$ can be characterized as the largest subgroup of G in which H is normal.

When (W, S) is a Coxeter system, the groups $C(H)$ and $N(H)$ are not known for arbitrary subgroups of W. (In fact, very little is known about the general subgroup structure of an arbitrary Coxeter group W.) However, often one can say a great deal more about these groups when H is a parabolic subgroup, $H = W_T$, $T \subseteq S$.

We first describe the normalizer of a given parabolic subgroup, and then turn our attention to the centralizer, which will prove more difficult to compute.

2.2.1 *The normalizer of a parabolic subgroup*

Let (W, S) be an arbitrary Coxeter system, and let W_T be a parabolic subgroup of W. (It is easy to see that we can reduce to the case of such standard parabolic subgroups, since $N(wHw^{-1}) = wN(H)w^{-1}$, for any $H \leq W$ and $w \in W$.) We first describe explicitly the elements $w \in W$ which conjugate one standard parabolic subgroup to another; it is a short step from there to being able to compute a generating set for the normalizer of W_T.

Let $T \subseteq S$, and consider $s \in S \setminus T$. We define $T(s)$ to be the subset of S which generates the (irreducible) direct factor of $W_{T \cup \{s\}}$ containing s. Suppose that $W_{T(s)}$ is spherical. By Exercise 1.7 both $W_{T(s)}$ and $W_{T(s) \setminus s}$ contain unique elements of greatest length (with respect to the generating set S). Denote these elements by $u(T, s)$ and $v(T, s)$, respectively, and define

$$w(T, s) = v(T, s)u(T, s). \tag{2.2}$$

The fact that both $u(T, s)$ and $v(T, s)$ are involutions, conjugation by which permutes the generators of the respective parabolic subgroups, leads to the following:

Lemma 2.1 *The element $w(T, s)$ defined as in (2.2) satisfies*

$$w(T, s)^{-1} T w(T, s) = (T \cup \{s\}) \setminus \{s_0\}$$

for some $s_0 \in T(s)$.

This lemma enables us to conjugate from one subset of S to another. We can illustrate this relationship by means of a directed, edge-labeled graph Γ. The vertices of Γ are the subsets of S, and there is a directed edge from T to T' (with label s) whenever the direct factor of $T \cup \{s\}$ containing s is spherical and $w(T, s)^{-1} T w(T, s) = T'$.

The computations above suggest that if there is a path in Γ from T to T', then W_T and $W_{T'}$ are conjugate to one another. This is the content of the next result, which utilizes the language of root systems introduced in 1.2.2.

Proposition 2.1 *Suppose T and T' are subsets of S, and suppose that $w \in W$ satisfies*

$$\{\alpha_t \mid t \in T\} = \{w\alpha_{t'} = \alpha_{wt'w^{-1}} \mid t' \in T'\}$$

where for any $s \in S$, α_s is the simple root corresponding to s. Then there is a directed path $T = T_0, T_1, ..., T_k = T'$ in Γ, with $[T_iT_{i+1}]$ bearing the label s_i $(0 \leq i \leq k - 1)$, such that

$$w = w(T_0, s_0) \cdot w(T_1, s_1) \cdots w(T_{k-1}, s_{k-1}).$$

Moreover, the expression given on the right-hand side of this equation is a geodesic representation for w.

The verification of this proposition is non-trivial, and we refer the interested reader to [Deodhar (1982)], where it was first proven.

From Proposition 2.1 follows the theorem we seek:

Theorem 2.4 *Let Γ be defined as above, and let $T \subseteq S$. Let $\mathcal{T}$ be a spanning tree in the connected component C of Γ containing the vertex T. For each vertex T' in $\mathcal{T}$ let $w_{T'}$ be the element of W determined by the unique path from T to T' as in Proposition 2.1. Then $N_W(W_T)$ is generated by T itself and all products of the form*

$$w_{T'} w(T', s) w_{T''}^{-1}$$

where there is a directed edge labeled s from T' to T'', $T'' \in C$.

Proof. The theorem follows almost immediately once we show that each $w \in N_W(W_T)$ corresponds (as above) to a closed path in C, since the generators given above are sufficient to generate the fundamental group of C. Suppose $w \in N_W(W_T)$. Left multiplying w by elements of W_T, we obtain a new element w' which satisfies the hypothesis of Proposition 2.1 with $T' = T$ (*cf.* 1.3.2). The rest of the proof is clear. $\qquad\square$

Remark 2.1   A more precise description of $N_W(W_T)$ is obtained by Brink and Howlett in [Brink and Howlett (1999)]. However, Theorem 2.4 will be sufficient for our purposes.

We will need two corollaries of the work above. A *two-dimensional* Coxeter system (W, S) is a system for which $\Sigma(W, S)$ is a two-dimensional complex. (Equivalently, every spherical subgroup of W has rank at most 2.)

Corollary 2.1   *Suppose (W, S) is a two-dimensional Coxeter system, and let $s, t \in S$, $s \neq t$. Then $N_W(W_{\{s,t\}}) = W_{\{s,t\}}$. Suppose furthermore that $w \in W$ satisfies $\{s, t\} = \{wsw^{-1}, wtw^{-1}\}$.*

1. *If $m = m_{st} > 2$ is even, then $w \in \{1, (st)^{m/2}\}$, and $wsw^{-1} = s$, $wtw^{-1} = t$.*
2. *If $m = m_{st}$ is odd, then $w \in \{1, (st)^{\frac{m-1}{2}} s\}$, and $w = 1 \Leftrightarrow wsw^{-1} = s \Leftrightarrow wtw^{-1} = t$.*

Proof.   All of the statements in the corollary follow from the fact that every vertex of Γ (defined as above) corresponding to a subset $T \subseteq S$ such that $|T| \geq 2$ is isolated. The details are left to the reader. $\square$

More careful analysis of spherical Coxeter groups and repeated application of Theorem 2.4 allows one to strengthen Corollary 2.1:

Corollary 2.2   *Suppose (W, S) is an arbitrary Coxeter system, and let $s, t, s', t' \in S$ such that $s \neq t$ and $s' \neq t'$. Then $W_{\{s,t\}}$ and $W_{\{s',t'\}}$ are conjugate to one another if and only if there is a sequence of edges $[st] = [s_0 t_0], [s_1 t_1], ..., [s_k t_k] = [s't']$, each labeled 3, such that for all i, $0 \leq i \leq k - 1$, the edges $[s_i t_i]$ and $[s_{i+1} t_{i+1}]$ form two of the sides of a triangle with edge label multiset $\{2, 3, 3\}$.*

2.2.2   *Centralizers in Coxeter groups*

Computing centralizers is often more difficult than computing normalizers. This is indeed the case with Coxeter groups. In what follows, we will describe completely the centralizer of the two-element group W_s ($s \in S$), as well as $C_W(wW_T w^{-1})$ for $wW_T w^{-1}$ a parabolic subgroup of an *even* system (W, S). (Although we will not make use of them here, methods used by H. Servatius in [Servatius (1989)] can be used to compute $C_W(w)$ for an arbitrary $w \in W$ when W is right angled. More general results than this are as yet unknown.)

Consider an arbitrary system (W, S) (with diagram $\mathcal{V}$), and let $s \in S$. Define $C(s) = C_W(W_s)$. We first describe $C(s)$ in terms of the generating set S, and then indicate two different ways that this characterization can be obtained.

Given the diagram $\mathcal{V}$, let $[st]$ be any edge in $\mathcal{V}$. If $m_{st} = 2k$ is even, define $u_{st} = (st)^{k-1}s$ (so that u_{st} commutes with t). If $m_{st} = 2k+1$ is odd, define $v_{st} = (st)^k$ (so that v_{st} conjugates s to t). It is easy to see that if $\{[ss_1], [s_1s_2], ..., [s_k t]\}$ is a path from s to t consisting of odd-labeled edges, then $v_{s_k t} v_{s_{k-1} s_k} \cdots v_{ss_1}$ conjugates s to t (*cf.* Proposition 5.3).

Let $\mathcal{V}_{\mathrm{odd}}$ be the diagram obtained from $\mathcal{V}$ by removing all edges with even labels. For $s \in S$, let $\mathcal{B}(s)$ be any collection of simple circuits in $\mathcal{V}_{\mathrm{odd}}$, each of which contains s, such that $\mathcal{B}(s)$ generates the fundamental group of the connected component of s in $\mathcal{V}_{\mathrm{odd}}$.

Theorem 2.5 *Let (W, S), $s \in S$, and $C(s)$ be as above. Then $C(s)$ is the subgroup of W generated by*

$$A \cup B \cup \{s\},$$

where A is given by

$$\left\{ vu_{ts_k}v^{-1} \mid v = v_{s_1 s} v_{s_2 s_1} \cdots v_{s_k s_{k-1}}; t, s_i \in S; m_{ts_k} \text{ even}; m_{s_1 s}, m_{s_i s_{i-1}} \text{ odd} \right\}$$

and B is given by

$$\left\{ v_{s_1 s} v_{s_2 s_1} \cdots v_{ss_k} \mid \{[ss_1], ..., [s_k s]\} \in \mathcal{B}(s) \right\}.$$

Roughly speaking, the elements of A correspond to "odd paths" based at s (with an "even spike" on the end), and the elements of B correspond to "odd loops" based at s. This is demonstrated in Figure 2.1, in which $s_1 s s_1 s_2 s_1 s_3 s_2 s_4 s_3 s_4 s_3 s_4 s_3 s_1 s_4 s s_1$ (solid arrows) corresponds to an odd loop and $s_1 s s_4 s_1 s_6 s_4 s_7 s_6 s_7 s_4 s_6 s_1 s_4 s s_1$ (hollow arrows) corresponds to an odd path with an even spike.

This description of $C(s)$ is a translation into the language of the generating set S of the result of Brink in [Brink (1996)], which gives a formula for $C(s)$ in terms of the group's action on the root system corresponding to (W, S). Brink's proof is carried out entirely in the context of the root system.

A more elementary proof can be obtained by judicious application of the van Kampen diagrams (as in Section 1.4). Indeed, the most difficult part of the proof is to show that if $wsw^{-1} = t$ holds for any $s, t \in S$, then

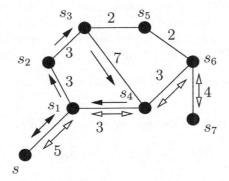

Fig. 2.1 The centralizer of s, as seen in $\mathcal{V}$

w can be written as a product of words $u_{s_i s_{i+1}}$ and $v_{s_i s_{i+1}}$ defined as above (perhaps with a letter s at the end).

Computing $C_W(wW_Tw^{-1})$ is more difficult when $|T| \geq 2$. We denote by $C(T)$ the group $C_W(W_T)$. The following theorem will be useful:

Theorem 2.6 *Suppose (W, S) is an even Coxeter system, and let $T \subseteq S$. Define*

$$A = S \cap C(T)$$

and

$$B = \left\{ (st)^k \mid s \in S, t \in T, m_{st} = 2k \ (k > 1), \text{ and } \{s,t\} \subseteq C(T \setminus \{s,t\}) \right\}.$$

Then $C(T)$ is generated by $A \cup B$.

The sets A and B consist of the elements of W which obviously commute with each generator $t \in T$; the theorem tells us that these elements are enough to generate $C(T)$.

We give a bare outline of a proof of Theorem 2.6. The first (and most difficult) step is to prove the theorem when $|T| = 2$, using van Kampen diagram arguments much like those used in Section 1.4. We then consider T such that $|T| \geq 3$. If there exists $t \in T$ such that $t \in C(T)$, then we can appeal to an inductive hypothesis, since in this case

$$C_W(T) = C_{C(t)}(T \setminus \{t\}).$$

When no such generator $t \in T$ is present, it is possible to decompose $C(T)$ as a direct product $Z(W_T) \times W_A$, where $Z(W_T)$ denotes the center of

W_T and A is as in Theorem 2.6. One quickly reduces to the case in which T spans a spherical simplex in $\mathcal{V}$, and it is straightforward to compute $Z(W_T)$ in this instance, and to show that it is generated by the set B defined above.

In practice, $C(T)$ will often be trivial. For instance, we have the

Corollary 2.3   *Suppose (W, S) is a two-dimensional even Coxeter system.*

1. *If $s \in S$, then $C(s)$ is generated by s and the collection*

$$\{(ts)^{k-1}t \mid m_{st} = 2k < \infty\}.$$

2. *If $s, t \in S$ such that $m_{st} = 2k < \infty$ (respectively, $m_{st} = \infty$), then $C(\{s, t\}) = \{1, (st)^k\}$ (respectively, $C(\{s, t\})$ is generated by $\mathrm{Lk}_2(s) \cap \mathrm{Lk}_2(t)$).*

3. *If $|T| \geq 3$, then $C(T)$ is trivial.*

2.3   Visual decompositions of Coxeter groups

In this section we give a brief introduction to the theory of graph-of-groups decompositions developed by J.-P. Serre and H. Bass, and indicate how this theory relates to Coxeter groups. A standard reference for this material in general is [Serre (1980)], although the first chapter of [Dicks and Dunwoody (1989)] gives an excellent overview. Our treatment of visual decompositions of Coxeter groups follows the work of M. Mihalik and S. Tschantz in [Mihalik amd Tschantz (preprint)].

2.3.1   *Graph-of-groups decompositions*

Two constructions frequently used to create new groups from old groups are free products with amalgamation and HNN-extensions.

Let G_1 and G_2 be groups with presentations

$$G_i \cong \langle S_i \mid R_i \rangle.$$

(We can assume the sets S_i are disjoint.) Let $H \leq G_1$ be a subgroup, and let $\phi : H \to G_2$ be a specified embedding of H into G_2. The *free product of G_1 and G_2, amalgamating over H* (denoted $G_1 *_H G_2$, where ϕ is understood) is the group G with presentation

$$\langle S_1, S_2 \mid R_1, R_2, h = \phi(h) \text{ for all } h \in H \rangle.$$

If $H = \{1\}$, $G_1 *_H G_2$ is obviously the free product $G_1 * G_2$ of G_1 and G_2.

Now let G be a group with presentation $\langle S \mid R \rangle$, and let $H \leq G$ be a subgroup, with $\phi : H \to G$ a specified embedding. Let t be a letter not appearing in S. The *HNN-extension of G by ϕ* (denoted $G *_{H,\phi} t$, or merely $G *_H t$ when ϕ is understood) is the group with presentation

$$\langle S, t \mid R, t^{-1}ht = \phi(h) \text{ for all } h \in H \rangle.$$

If $H = \{1\}$, then $G *_H t \cong G * \mathbb{Z}$. If $H = G$ and the embedding ϕ is the identity, then $G *_H t \cong G \times \mathbb{Z}$.

These constructions can be iterated, applying one after another. This iterated procedure can be diagrammed nicely by means of a graph of groups.

Let Γ be a connected graph, with $V = V(\Gamma)$ its set of vertices and $E = E(\Gamma)$ its set of oriented edges. Let $\iota, \tau : E \to V$ be the maps that associate to each edge $e \in E$ the initial and terminal vertices $\iota(e)$ and $\tau(e)$ of that edge, respectively. To each $x \in V \cup E$, we associate a group, G_x, such that

1. $G_e \subseteq G_{\iota(e)}$ for all $e \in E$, and
2. there is an injective homomorphism $t_e : G_e \to G_{\tau(e)}$ for all $e \in E$.

The image of $g \in G_e$ under t_e is frequently denoted g^{t_e}.

Such a structure (consisting of Γ itself, all of the groups G_x, as well as the embeddings t_e) is called a *graph of groups*. We now define a presentation for the *fundamental group* $\pi_1(\Gamma)$ of the graph of groups Γ. The generating set S for π_1 is the disjoint union of the vertex groups G_v ($v \in V$), along with a new symbol t_e for each edge $e \in E$. The relations consist of defining relations for every vertex group G_v ($v \in V$), the relation $g t_e = t_e g^{t_e}$ for every $g \in G_e$ and $e \in E$, and $t_e = 1$ for every edge of an arbitrary but fixed spanning tree for Γ.

Remark 2.2 Clearly this definition generalizes the fundamental group of an unadorned graph Γ, which can be considered a graph of groups in which every group G_x is trivial, $x \in V \cup E$. It is also clear that $G_1 *_H G_2 \cong \pi_1(\Gamma)$ when Γ consists of a single edge $e = [v_1 v_2]$, $G_{v_i} \cong G_i$, $G_e \cong H$, and $t_e(h) = \phi(h)$ for all $h \in H$, where $\phi : H \to G_{v_2}$ is as before. Likewise, $G *_H t \cong \pi_1(\Gamma)$ when Γ consists of a single vertex v with an edge e such that $\iota(e) = \tau(e) = v$, $G_v \cong G$, $G_e \cong H$, and t_e is the embedding specified in the definition of $G *_H t$. More generally, the fundamental group $\pi_1(\Gamma)$ represents the result of applying a number of free products with amalgamation and HNN-extensions in sequence.

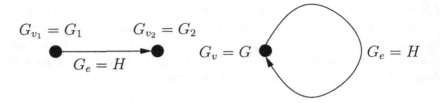

Fig. 2.2 Graphs of groups for a free product with amalgamation and an HNN-extension

2.3.2 *Visual decompositions*

As a quick survey of the current research literature will demonstrate, a great deal is known about the structure of groups formed by applying free products with amalgamation and HNN-extension to groups whose structures are already known. (For a concise summary of a number of results, consult [Lyndon and Schupp (1977)].) Therefore it is often desirable to express a given group G as the fundamental group of a graph of groups. In case (W, S) is a Coxeter system, this can be done quite easily, simply by looking at the diagram $\mathcal{V}$ for (W, S). Moreover, such "visual" decompositions turn out to be canonical in a well-defined sense.

As a straightforward example, suppose that $S = S_1 \cup S_2$, and denote by T the intersection $S_1 \cap S_2$ (which could be empty). Suppose further that every edge in the diagram $\mathcal{V}$ is contained in one of the full subdiagrams of $\mathcal{V}$ induced by S_1 and S_2. It is not hard to see that $W \cong W_{S_1} *_{W_T} W_{S_2}$. This is an example of a *visual (graph of groups) decomposition* of the Coxeter group W. We now make this notion precise.

Let (W, S) be given, with diagram $\mathcal{V}$. A graph of groups Γ is said to be a visual decomposition of W if $W \cong \pi_1(\Gamma)$ and the following three conditions hold:

1. every group G_x ($x \in V \cup E$) is a standard parabolic subgroup W_T for some $T \subseteq S$,
2. every edge in $\mathcal{V}$ is contained in the full subdiagram of $\mathcal{V}$ induced by some vertex group G_v ($v \in V$), and
3. for every $s \in S$, the set of $x \in V \cup E$ such that $s \in G_x$ is a nonempty subtree of Γ.

Remark 2.3 If Γ is a graph of groups for which the underlying graph is not simply connected, then the fundamental group of Γ will map homorphically onto $\mathbb{Z}$. (Indeed, for some edge $e \in E$ not contained in a fixed spanning tree, map t_e to $1 \in \mathbb{Z}$, and map every other generator to 0.) But every homomorphism from a Coxeter group W to $\mathbb{Z}$ must map each generator $s \in S$ to 0; thus W cannot be the fundamental group of a graph of groups which is not a tree, and W can be expressed only as the result of iterated free products with amalgamation.

Clearly a given group W may decompose in a number of different ways; we will be interested primarily in fairly straightforward decompositions (for instance, those in which the group W_T is spherical, or consists only of a single edge, or a single vertex). The following theorem (the main theorem from [Mihalik amd Tschantz (preprint)]) indicates the usefulness of visual decompositions.

Theorem 2.7 *Let (W, S) be a Coxeter system with diagram $\mathcal{V}$, and let W be a subgroup of $\pi_1(\Gamma)$ for the graph of groups Γ. Then W admits a visual decomposition $\Gamma(W)$ where each vertex group of $\Gamma(W)$ is a subgroup of a conjugate of a vertex group of Γ and each edge group of $\Gamma(W)$ is a subgroup of a conjugate of an edge group of Γ. Furthermore, $\Gamma(W)$ can be taken so that each standard parabolic subgroup of W that is a subgroup of a conjugate of a vertex group of Γ is a subgroup of a vertex group of $\Gamma(W)$.*

In some sense, then, *every* graph of groups decomposition of W is nearly a visual decomposition.

Exercises 9-11 will ask the reader to investigate the way in which this theorem interacts with *diagram twists*, defined in 3.2.2. We will draw upon this interaction in Chapter 6.

2.4   The Word and Conjugacy Problems

Before moving on to address the Isomorphism Problem (the underlying focus of the remainder of the text), let us take a moment to examine the other fundamental problems of combinatorial group theory, *vis-à-vis* Coxeter groups.

Recall that the *Word Problem* is said to be solvable for a finitely presented group if there exists an algorithm which determines whether a word in the given generating set for the group represents the identity element of the group. (By extension, we may speak of the solvability of the Word Problem in a particular class of groups.) The *Conjugacy Problem* is said to be solvable for a group if there exists an algorithm which determines whether two words in the given generating set for the group represent conjugate elements in the group. This is clearly a harder problem than the Word Problem.

Theorem 2.8   *Both the Word and Conjugacy Problems are solvable in the class of Coxeter groups.*

We briefly indicate a few proofs of the Word Problem, and a single proof of the Conjugacy Problem.

We have already touched upon a characterization of Coxeter groups which leads to one solution to the Word Problem. Namely, we have seen that any Coxeter group can be considered as a group of linear transformations on some finite-dimensional vector space. Thus Coxeter groups are linear groups. It is known that all linear groups have solvable Word Problem.

This proof is rather disappointing, as it doesn't give us any further insight into the structure of the groups. Tits (in [Tits (1969)]) gave a more satisfying proof of the Word Problem's solvability by providing an explicit algorithm for determining whether a given word represents the identity or not. We now explain this algorithm.

Given a word w in the generating set S, we define a set, $D(w)$, of words which can be derived from w by applying any sequence of operations of the following two types:

(D1) For any $s \neq t$ in S such that $m_{st} < \infty$, replace any occurrence in w of the word $stst \cdots$ of length m_{st} with the word $tsts \cdots$ of length m_{st}.
(D2) Cancel any adjacent occurrences of the same letter $s \in S$.

It is clear that $D(w)$ is finite, for any given w, and that it can be computed quite efficiently. In [Tits (1969)] we find the following result:

Proposition 2.2   *Let w, w' be words in the generating set S for the Coxeter group W. Then w is reduced if and only if every element of $D(w)$ has the same length (in the free group on S) as w. Moreover, w and w' represent the same element of W if and only if $D(w) \cap D(w') \neq \emptyset$.*

We thus obtain a straightforward algorithm which solves the Word Problem for Coxeter groups. The following restatement of part of Tits's solution will be prove very useful later:

Proposition 2.3   *Suppose w is a word in the generating set S which contains no subword $stst \cdots$ of length $m_{st} < \infty$. Then w is reduced.*

Now for one final solution to the Word Problem; this one goes hand-in-hand with a solution to the Conjugacy Problem.

Theorem 2.9   *Let G be a group which acts properly and cocompactly by isometries on a* CAT(0) *metric space, X. Then G is finitely presented, and G has solvable Word and Conjugacy Problems.*

We sketch a proof, following [Bridson and Haefliger 1999], III.Γ.1.

Proof.   First, we use the cocompactness of the action of G in order to find a useful generating set for G. Namely, let $d > 0$ such that

$$X = \bigcup_{g \in G} g \cdot B(x_0, \frac{d}{3})$$

for some fixed point $x_0 \in X$. ($B(x, r)$ is the ball in X of radius r around the point x.) Then let

$$S = S(d) = \{g \in G \mid d(g \cdot x_0, x_0) \le d + 1\},$$

where $d(\cdot, \cdot)$ is the metric with respect to which X is CAT(0). It can be shown that this is a generating set for G.

Now let $R = R(d)$ be the set of reduced (with respect to S) words of S-length at most 10 in the letters $S^{\pm}$ which are equal to the identity in G. We claim that now a word in the letters $S^{\pm 1}$ represents the identity of G if and only if w is freely equal to a product of the form

$$\prod_{i=1}^{n} x_i r_i x_i^{-1}$$

where $n \le (d + 1)|w|^2$, $r_i \in R$ for all i, and $|x_i| \le (d + 1)|w|$ for all i. (Here, $|x|$ is the length of the word x with respect to the generating set S.) The solvability of the word problem follows from this fact, since clearly there are only finitely many such products, for any word of a given length. Furthermore, it is easy to compute the time complexity of this algorithm. (What is it?)

To prove the above claim, we make more use of the geometry of the space X. Given an element g of G, we find a word g' in the generators S such that $g =_G g'$. This word g' is a product $s_1 \cdots s_k$ of letters $s_i \in S$, each of which is a ratio of words which approximate the unique geodesic path in X from x_0 to $g \cdot x_0$. (The defining property of S plays a major role here.) Furthermore, g' is defined in such a way that for all $s \in S$, the ratio

$g's[(gs)']^{-1}$ is freely equal to a product

$$\prod p_i r_i p_i^{-1}$$

for some $r_i \in R$, and in which the number of terms in the product is independent of the choices of g and s. (The proof of this fact requires the convexity of the CAT(0) metric.) Finally, if $w = s_1 \cdots s_m =_G 1$, we may express w as a product of ratios $g's[(gs)']^{-1}$ as above, and therefore as a product of conjugates of elements of R. Careful bookkeeping yields an upper bound on the number of terms which depends only on m, the S-length of w.

Our solution to the Conjugacy Problem uses the solution to the Word Problem, as well as the fact that groups acting nicely on CAT(0) spaces satisfy the *quasi-monotone conjugacy (q.m.c.) property*. G is said to have this property if there exists a constant $k < \infty$ depending only on G such that if u and v are conjugate, then there is a word w satisfying $wuw^{-1} = v$ and $|w| \leq k \cdot \max\{|u|, |v|\}$. That q.m.c. holds in a group acting nicely on a CAT(0) space X follows from the convexity of the metric with respect to which X is CAT(0).

Assume that the q.m.c. property has been established for G, and that G has solvable Word Problem. Given $n < \infty$, we can define a graph $\Gamma(n)$ whose vertices are freely reduced words w of S-length at most n over the alphabet $S^{\pm}$, and in which there is an edge $[w_1 w_2]$ if and only if w_1 and w_2 are conjugate by an element of s. (This conjugacy can be checked by using the solution to the Word Problem.) Then any two words u and v are conjugate if and only if there is some path in $\Gamma(k \cdot \max\{|u|, |v|\})$, where k is the constant of quasi-monotonicity.

We note that this algorithm is very inefficient (roughly speaking, it is exponential in the length of the words u and v).   □

It now follows immediately from Theorems 1.4 and 2.9 that Coxeter groups have solvable Conjugacy Problem (finite presentability and solvability of the Word Problem being known already).

Theorem 2.9 demonstrates yet again the usefulness of knowing that W acts nicely on a well-behaved space. (*Cf.* Lemma 1.2.) The analysis of such group actions is one of the cornerstones of the field of geometric group theory, and the literature on this topic is incredibly rich. The reader is urged to consult [Bridson and Haefliger 1999] for a more in-depth look at group actions on CAT(0) spaces, as well as a plethora of other notions in geometric group theory.

2.5 Exercises

1. Compute the homology of the free group F_n on n generators by using a $K(G, 1)$-complex for F_n.

2. Prove Corollary 2.2. (Recall Lemma 1.1.)

3. The Coxeter system (W, S) is called *locally reducible* if every spherical subgroup is isomorphic to a direct product of dihedral groups D_k and the 2-element group $\mathbb{Z}_2$. Using Corollary 2.2, prove that no two distinct rank-2 standard parabolic subgroups of a locally reducible system lie in the same conjugacy class.

4. Prove all of the statements made in Remark 2.2.

5. Prove that the Word Problem is solvable for finitely presented abelian groups by giving an explicit algorithm which decides whether or not an input word represents the trivial element. What is the time complexity of the algorithm you gave? (That is, what is the nature of the relation which gives the length of time required by the algorithm as a function of the length of the input word? For instance, is it a linear function? A quadratic function?)

6. Give another proof that finitely presented abelian groups have solvable Word Problem by demonstrating, for such a given group G, a CAT(0) space upon which G acts properly and cocompactly by isometries.

7. Prove that the Word Problem is solvable for the so-called *Baumslag-Solitar group* $B(1, 2) = \langle a, t \mid tat^{-1} = a^2 \rangle$ by giving an explicit algorithm. What is the time complexity of the algorithm you gave?

8. Suppose that (W, S) is an even Coxeter system, as defined in 1.2.1. Let $s, t \in S$ and suppose that st has finite order $m > 2$. If $w \in W$ satisfies $wststw^{-1} = (stst)^{\pm 1}$, prove that w can be written geodesically as uv, where $u \in W_{\{s,t\}}$ and v is a single generator commuting with both s and t. (*Hint:* use Theorem 2.6 and Exercise 1.9, where the cosets used in the latter are taken with respect to the subgroup $W_{\{s,t\}}$.)

The next three exercises explore results first proven in [Mühlherr and Weidmann (2002)]. Completion of these exercises requires the definition of a *diagram twist*, given in 3.2.2.

9. Let (W, S) be a Coxeter system. Suppose that W can be visually decomposed as a product $A *_C B$, for $A = W_{S_1}$, $B = W_{S_2}$, and $C = W_{S_1 \cap S_2}$,

for $S = S_1 \cup S_2$. Suppose further that $|C| < \infty$, and let Δ be its unique longest element.

a. Show that the groups $A' = A$, $B' = \Delta B \Delta^{-1}$, and $C' = A' \cap B'$ yield a new visual decomposition of W, relative to a new fundamental generating set. Do the same for $A'' = \Delta^{-1} A \Delta$, $B'' = B$, and $C'' = A'' \cap B''$.

b. Prove that the diagrams corresponding to the two visual decompositions given above are isomorphic as labeled graphs.

Define fundamental generating sets S' and S'' to be *twist equivalent* to one another if one can be obtained from the other by applying a (clearly finite) sequence of operations as in part a.

c. Suppose S' and S'' are twist equivalent fundamental generating sets. Prove that if $\{s', t'\} \subseteq S'$ gets replaced with $\{s'', t''\}$ in passing from S' to S'', then the dihedral groups generated by these sets are conjugate in W.

10. Suppose that (W, S) is a Coxeter system yielding a visual decomposition $A_1 *_{C_1} A_2 *_{C_2} A_3$, where $C_1 = w C_2 w^{-1}$ ($w \in A_2$) is a finite group. Use Proposition 2.1 to prove that W has a visual decomposition $A_1 *_{C_1} A_2 *_{C_1} w A_3 w^{-1}$ with respect to some fundamental generating set S'.

11. Let (W, S) be a Coxeter system, and let C be a spherical subgroup such that W does not admit a free product decomposition by amalgamating over any proper subgroup of C. Use the previous exercise and Theorem 2.7 to prove that W admits an amalgamated product decomposition of the form

$$W = A_1 *_C A_2 *_C \cdots *_C A_k$$

such that this decomposition is visual with respect to some fundamental generating set S' (S' and S twist equivalent) and no A_i admits an amalgamated free product decomposition over a proper subgroup of C. (*Hint:* begin with the "coarsest" decomposition possible, and refine it successively by splitting terms in the decomposition whenever necessary.)

Chapter 3

Rigidity

We now introduce the concept of rigidity, along with a number of other characterizations of "uniqueness" of Coxeter group presentations. After a brief examination of the connections between these concepts, we will give a survey of the results with which the remainder of the text will deal. The number of such results is quite large; the reader ought not feel overwhelmed by their sudden and superficial introduction in this chapter! Rather, careful analysis of the statements of the theorems below may allow the reader to detect the flavor of the proofs which will come in later chapters. Therefore the reader may wish to view this chapter as a "book within a book", wherein the rest of the text is succinctly summarized.

3.1 Rigidity and related conditions

Let us begin with some definitions.

3.1.1 *Rigidity and strong rigidity*

Let W be a Coxeter group. We say that W is *rigid* if given any two systems (W, S) and (W, S') for W, there is an automorphism $\alpha \in \text{Aut}(W)$ satisfying $\alpha(S) = S'$. Equivalently, W is rigid if given any two diagrams $\mathcal{V}$ and $\mathcal{V}'$ for the group W, $\mathcal{V}$ and $\mathcal{V}'$ are isomorphic as edge-labeled graphs. (That is, there exists a map γ from the vertex set of $\mathcal{V}$ to the vertex set of $\mathcal{V}'$ such that $[st]$ is an edge labeled n in $\mathcal{V}$ if and only if $[\gamma(s)\gamma(t)]$ is an edge labeled n in $\mathcal{V}'$.) We can paraphrase this condition by saying that W is rigid if and only if the system (W, S) is determined by the group W, up to automorphism.

Immediately we see that there are groups which are not rigid; in particular, if k is odd, the dihedral group D_{2k} is not rigid. Moreover, all other dihedral groups are rigid, so that there exist groups with this property. Thus it is not a trivial question to ask whether or not a given Coxeter group is rigid.

We may strengthen this condition a bit. We call the Coxeter group W *strongly rigid* if given any two systems (W, S) and (W, S') for W, there exists an *inner* automorphism $\alpha \in \mathrm{Inn}(W)$ satisfying $\alpha(S) = S'$. Clearly strong rigidity implies rigidity, so in particular any two diagrams for a strongly rigid group W are isomorphic as edge-labeled graphs.

It is also clear that there are groups which are rigid but not strongly rigid, so this definition too serves a purpose. In fact, consider the group $W = \mathbb{Z}_2 \times \mathbb{Z}_2 \cong \langle a, b \mid a^2, b^2, (ab)^2 \rangle$. Letting $c = ab$, W can be presented $\langle a, c \mid a^2, c^2, (ac)^2 \rangle$, which clearly yields a diagram isomorphic to the first, but the element c is a conjugate of neither a nor b.

We will return throughout this text to the relationships between rigidity and strong rigidity in a number of contexts.

3.1.2 *Reflection independence*

Because of the strong geometric flavor of Coxeter group theory, it should come as no surprise that the structure of a given Coxeter system is intimately related to its set of reflections. Indeed, as we will see in later chapters, two systems (W, S) and (W, S') which yield the same sets of reflections often exhibit much greater similarity than two systems which do not yield the same sets of reflections. This fact motivates the following definition.

Let W be a Coxeter group. We say that W is *reflection independent* if any two systems (W, S) and (W, S') for W yield the same reflections. That is,

$$\{wsw^{-1} \mid s \in S, w \in W\} = \{ws'w^{-1} \mid s' \in S', w \in W\}. \qquad (3.1)$$

First note that there are groups which are not reflection independent: the group $\mathbb{Z}_2 \times \mathbb{Z}_2$ is such a group (see 3.1.1). It is also clear that some groups are reflection independent. For instance, if W is strongly rigid, then W is also reflection independent. The converse is not true, however, as is demonstrated by the following group:

$$\langle a, b, c, d \mid a^2, b^2, c^2, d^2, (ab)^4, (bc)^4, (ac)^4, (ad)^4 \rangle. \qquad (3.2)$$

(That this group is reflection independent but not strongly rigid will follow from Theorems 3.3 and 3.12, though the reader invited to supply his or her own proof at this time.)

Somewhat closely related to the concept of reflection independence is that of reflection preservation. Let (W, S) be a Coxeter system and let $\alpha \in \text{Aut}(W)$. It is easy to show that $\alpha(S)$ is also a fundamental generating set, yielding a new system $(W, \alpha(S))$. We will say that the system (W, S) is *reflection preserving* if for every automorphism $\alpha \in \text{Aut}(W)$, the system $(W, \alpha(S))$ yields the same reflections as (W, S).

If W is reflection independent, then (W, S) is reflection preserving for every fundamental generating set S. The converse, however, is not true, as is shown by the dihedral group D_6.

3.1.3 *Reflection rigidity and strong reflection rigidity*

We now define two "uniqueness" conditions which are slight weakenings of the notions of rigidity and strong rigidity, respectively.

We say that a given Coxeter system (W, S) is *reflection rigid* if for any other system (W, S') which yields the same reflections as (W, S), there is an automorphism $\alpha \in \text{Aut}(W)$ such that $\alpha(S) = S'$. That is, W is rigid when we restrict our attention to the class of Coxeter systems which yield the same reflections. (We may paraphrase reflection rigidity thus: (W, S) is reflection rigid if and only if the group W and its set of reflections R together determine the system (W, S), up to automorphism.)

Finally, we say that a given Coxeter system (W, S) is *strongly reflection rigid* if for any other system (W, S') which yields the same reflections as (W, S), there is an *inner* automorphism $\alpha \in \text{Inn}(W)$ satisfying $\alpha(S) = S'$.

Are these definitions necessary? That is, are there groups which are (strongly) reflection rigid but not (strongly) rigid? In fact, we will see that the ever-useful examples D_{2k} (k odd) provide a collection of groups which are reflection rigid but not (as shown in 3.1.1) rigid.

Note also that in case W is reflection independent, then (strong) reflection rigidity and (strong) rigidity are equivalent.

3.1.4 *Classes of Coxeter groups*

Before we mention specific theorems regarding rigidity and its variations, we motivate the ensuing discussion by giving an overview of the classes of Coxeter groups commonly considered below.

The simplest of the infinite Coxeter groups are the right-angled Coxeter groups. Because the structure of these groups is easier to understand than that of more general Coxeter groups, many facts about Coxeter groups are proven for right-angled groups first.

For reasons which will become apparent later, even Coxeter groups often provide a reasonable generalization of the right-angled groups. At the other extreme, one may also consider the *odd* Coxeter groups, those which have a presentation (W, S) whose corresponding diagram contains only edges with odd labels.

To apply some methods of proof (particularly small cancellation theory and some of the "edge matching" arguments which we will see in the following chapters), it is convenient to consider groups whose diagrams have only large edge labels. Therefore the classes of large-type and extra-large type Coxeter systems are often considered. Such systems have the property that every spherical subgroup has rank at most 2, facilitating arguments which make use of the subgroup structure of W. (Recall that the systems (W, S) for which every spherical subgroup has rank at most 2 are called *two-dimensional*. The defining condition is equivalent to the two-dimensionality of the Davis complex $\Sigma(W, S)$. We will pay close attention to these systems in the sequel.)

A system (W, S) is called *locally reducible* if every spherical subgroup W_T is isomorphic to a direct product of dihedral groups D_n and the two-element group $\mathbb{Z}_2$. (Thus, in particular, two-dimensional Coxeter systems are locally reducible.) We will see that local reducibility does not depend on the system, so that we may speak of a *group* being locally reducible. Like groups corresponding to large-type systems, locally reducible groups have rather straightforward subgroup structure, which sometimes makes their analysis easier. Moreover, the insight gleaned from proofs which address large-type groups may point the way toward extending these proofs to locally reducible groups, and beyond, to Coxeter groups in general.

Throughout our studies we will encounter other classes of Coxeter groups that will be defined in a more *ad hoc* fashion in order to fit the demands of a particular proof.

3.2 Some first results

We now state some early theorems regarding rigidity. These results are grouped together here not because their proofs are similar (they are not!), but merely because they were the first results proven, representing the van of the investigation of rigidity. Because of this, both the questions these results pose and the answers they provide serve to direct later study on the topic. Moreover, many of the methods used to prove the early theorems have been useful in more general contexts.

3.2.1 *Right-angled Coxeter groups*

The relatively simple structure of right-angled Coxeter groups has allowed more in-depth analysis of these groups. In particular, as we will see immediately, the Isomorphism Problem for right-angled groups is a trivial one. Moreover, the automorphism group $\text{Aut}(W)$ is completely understood when W is right-angled (see 3.5.1).

One of the first rigidity results to be proven was the following theorem, due to D. Radcliffe.

Theorem 3.1 *Let W be a right-angled Coxeter group. Then W is rigid.*

In particular, if one system for W is right-angled, then so is any other system.

Radcliffe's proof makes use of the fundamental fact (Lemma 1.2) that a finite group of isometries of a complete, simply connected CAT(0) space fixes some point of the space. Having already proven this fact, the remainder of the proof turns out to be largely combinatorial.

Using similar methods, Radcliffe was able to extend his results somewhat to a broader class of groups which behave very much like right-angled groups.

Theorem 3.2 *Let (W, S) be a Coxeter system such that every edge in the corresponding diagram $\mathcal{V}$ is labeled by either 2 or by an integer divisible by 4. Then W is rigid.*

The seemingly strange divisibility criterion arises in another early result (from [Bahls (2003)]), this one dealing with reflection independence:

Theorem 3.3 *Let (W, S) be a Coxeter system such that the label of any edge in the corresponding diagram $\mathcal{V}$ is either odd or divisible by 4. Then W is reflection independent.*

The divisibility criterion is significant for two obvious reasons. First, as mentioned in 3.1.4, even Coxeter groups are often easier to work with than arbitrary groups. Second, whereas dihedral groups of the form D_{2k} for k odd are not rigid, the dihedral groups of the form D_{4k} are.

To end this subsection, we mention our first result concerning strong rigidity, obtained in [Brady, *et al.* (2002)]. Theorem 3.4 characterizes strongly (reflection) rigid right-angled Coxeter groups.

Theorem 3.4   *Let (W, S) be a right-angled Coxeter system with diagram $\mathcal{V}$. Then (W, S) is strongly reflection rigid if and only if the following condition holds:*

1. *for every vertex s in $\mathcal{V}$ the full subgraph on the set of vertices not adjacent (or equal) to s is connected.*

Moreover, W is strongly rigid if and only if the following condition also holds:

2. *every vertex s in $\mathcal{V}$ is the intersection of all maximal simplices containing it.*

We will formulate partial generalizations of this result in the following sections. Even in the more general results, the structure imposed upon $\mathcal{V}$ by the conditions in Theorem 3.4 will still be apparent.

3.2.2   *The role of reflections*

As mentioned above, the presentation of a Coxeter group may depend heavily on the set of reflections which corresponds to the system (recall the case of the dihedral groups of order $4k$ for k odd). Let W be a Coxeter group. By considering only systems (W, S) and (W, S') which yield the same set of reflections, it is often possible to state rigidity results more cleanly.

Theorem 3.5 summarizes the primary results from [Brady, *et al.* (2002)]; in order to state this theorem, we must introduce the important notion of a *diagram twist*, first defined in [Brady, *et al.* (2002)].

Let (W, S) be an arbitrary Coxeter system with diagram $\mathcal{V}$, and suppose that S contains nonempty sets T_1 and T_2 satisfying the following conditions:

1. T_2 is spherical, and
2. for any $t_1 \in T_1$ and $t \in S \setminus (T_1 \cup T_2)$ such that $[t_1 t]$ is an edge in $\mathcal{V}$, $[t_2 t]$ is an edge labeled 2 for every $t_2 \in T_2$.

Let $\Delta = \Delta_{W_{T_2}}$ be the unique longest element in W_{T_2} (see Exercise 1.7). Replacing each $t_1 \in T_1$ by $\Delta t_1 \Delta$ and leaving every other generator of S unchanged yields a new generating set, S', for W. We claim that this is the fundamental generating set for a Coxeter system, (W, S'). In fact, if we modify $\mathcal{V}$ by replacing each edge $[t_1 t_2]$ ($t_1 \in T_1$, $t_2 \in T_2$) with the edge $[t_1 t_2']$, where $t_2' = \Delta t_2 \Delta$, the diagram $\mathcal{V}'$ which results will correspond to the system (W, S') (see Exercise 6). This operation is called a *diagram twist* around the set T_2. Figure 3.1 illustrates this concept with a simple example.

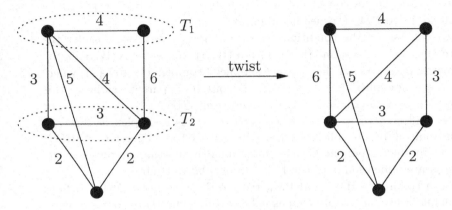

Fig. 3.1   A diagram twist

Theorem 3.5   *Let (W, S) be a Coxeter system with diagram $\mathcal{V}$. If either of the following conditions is true, then (W, S) is reflection rigid:*

1. *W is finite, or*
2. *(W, S) is even.*

If instead $\mathcal{V}$ is a tree, then (W, S) is reflection rigid, up to diagram twisting. (That is, given any other system (W, S') satisfying $R(S) = R(S')$, the diagram $\mathcal{V}'$ can be obtained from $\mathcal{V}$ by a series of diagram twists.)

Further results concerning (strong) reflection rigidity will be given in Section 3.4, and we will discuss the rigidity of a related class of groups in 3.5.3.

Using very different techniques, R. Charney and M. Davis were able to prove the following theorem, which relates the rigidity of a Coxeter group W to its potential to act nicely on an appropriate topological space.

Theorem 3.6   *Suppose that a Coxeter group W acts effectively, properly, and cocompactly on a contractible manifold. Then W is strongly rigid.*

Theorem 3.6 is included at this point because a key step in the proof of the theorem is showing that a group W admitting such an action is reflection independent (Proposition 4.4). The proof makes considerable use of the structure of the complex Σ defined in Section 1.5.3. In particular, two different classes of Coxeter groups for which Theorem 3.6 will hold are defined based upon the "homological" structure that the group W imposes upon Σ. The classes considered are the groups of types HM_n ("homology manifold") and PM_n ("psuedomanifold").

To understand the conditions HM_n and PM_n, we must consider the nerve of a Coxeter system (W, S). Given (W, S), the *nerve* $N(W, S)$ is the geometric realization of the abstract simplicial complex $(\uparrow_{\mathcal{S}} \emptyset) \setminus \{\emptyset\}$, where $\mathcal{S}$ is the set of spherical subsets of S. Despite its apparent nastiness, this complex is easy to visualize: there is a vertex in $N(W, S)$ for each generator $s \in S$, and a set $T = \{s_1, ..., s_k\}$ of vertices spans a simplex in $N(W, S)$ if and only if W_T is spherical. (See Exercise 1.13.) Therefore $N(W, S)$ can be obtained from the Coxeter diagram $\mathcal{V}$ by pasting simplices of the appropriate dimension onto the 1-skeleta of spherical subsets.

Both conditions HM_n and PM_n require that the nerve $N(W, S)$ have a "manifold-like" structure that is spherical in some (homological) sense. These conditions will be made more precise in Chapter 4. The hypothesis of Theorem 3.6 will then be equivalent to the statement that W is of type HM_n. The theorem will also be true of Coxeter groups of type PM_n. A proof of Theorem 3.6 in case (W, S) is of type HM_n will be given in Chapter 4. Moreover, as a consequence of this theorem, we will obtain information about the automorphism group $\mathrm{Aut}(W)$:

Corollary 3.1 *Let W be of type HM_n or PM_n. Then $\mathrm{Aut}(W)$ is a semidirect product of $\mathrm{Diag}(W)$ by $\mathrm{Inn}(W)$.*

Here, $\mathrm{Diag}(W)$ is the group of *diagram automorphisms*, those automorphisms of W induced by symmetries of the diagram $\mathcal{V}$ in the obvious fashion. This corollary, and generalizations of it, will be addressed in Chapter 7, when questions related to rigidity are discussed.

We round out our discussion of early results with another theorem which depends upon reflection independence.

In his dissertation, A. Kaul proves Theorem 3.7, which deals with Coxeter systems of type K_n. The system (W, S) is said to be of *type K_n* if its diagram $\mathcal{V}$ is a complete graph on n vertices for which every edge has an odd label. (Clearly all such groups are two-dimensional.)

Theorem 3.7 *Suppose that (W, S) is of type K_n and that furthermore all (or all but one) of the edges of $\mathcal{V}$ bears the same label. Then W is rigid.*

A key ingredient in Kaul's proof of Theorem 3.7 is the following proposition, whose usefulness justifies its mention sooner rather than later.

Proposition 3.1 *Suppose that (W, S) is a Coxeter system (with diagram $\mathcal{V}$) such that every maximal spherical subgroup of W has the same rank, and let (W, S') be another system for W (with diagram $\mathcal{V}'$). Then there is a one-to-one correspondence ϕ between the maximal spherical simplices of $\mathcal{V}$ and those of $\mathcal{V}'$ such that for every maximal spherical simplex $\sigma \subseteq \mathcal{V}$, there is a group element $w \in W$ satisfying $W_\sigma = w W_{\phi(\sigma)} w^{-1}$.*

Proposition 3.1 can be generalized to all Coxeter groups, as we will see in Section 4.1. In a sense, this proposition gives "local" information about the structure of the Coxeter system (W, S); this information will be used in order to understand the "global" structure of the entire group. Passing from local information to global information will be a recurrent theme in this volume.

3.3 Even Coxeter groups

The previous section demonstrates the diversity of methods that can be used to answer questions concerning rigidity. In this section, we highlight the point of view which considers the combinatorial goings-on within a given Coxeter group. In particular, our focus falls upon the Coxeter presentation itself, rather than on the group's geometric structure or its action upon a particular space. Throughout this section we suppose that the Coxeter system (W, S) is even.

Why study even Coxeter systems? To begin with, the subgroup structure of an even system is very easily understood: every spherical subgroup is isomorphic to a direct product of dihedral groups and copies of the two element group $\mathbb{Z}_2$ (that is, they are locally reducible; see Section 5.2). Moreover, there are a number of techniques that can be applied more easily to even systems than to arbitrary systems. For instance, the following fact will play an important role in the development of Chapter 5:

Proposition 3.2 *Let (W, S) be an even Coxeter system, and let $T \subseteq S$. Then there is a retraction $\rho : W \to W_T$. (The map ρ is obtained by identifying every generator $s \in S \setminus T$ to 1.)*

This proposition fails in the general case (see Exercise 4).

In Chapter 5 we will establish a correspondence between the spherical simplices of two even systems (W, S) and (W, S') which extends Proposition 3.1 in a natural fashion. The information gleaned from this correspondence will be the primary tool used in proving our next theorem.

In order to state Theorem 3.8, we need some more terminology. Given a Coxeter diagram $\mathcal{V}$ and a vertex s in $\mathcal{V}$, we define the *link* $\text{Lk}(s)$ of s to be the set $\{t \in S \mid [st]$ is an edge of $\mathcal{V}\}$. Define the *star* $\text{St}(s)$ of s by $\text{St}(s) = \text{Lk}(s) \cup \{s\}$. Define the 2-*link* $\text{Lk}_2(s)$ of s to be the set $\{t \in S \mid [st]$ is an edge of $\mathcal{V}$ labeled 2$\}$, and the 2-*star* $\text{St}_2(s)$ of s by $\text{St}_2(s) = \text{Lk}_2(s) \cup \{s\}$.

Theorem 3.8   *Let (W, S) be an even Coxeter system, with diagram $\mathcal{V}$.*

1. *If there are distinct vertices s and t in $\mathcal{V}$ such that $\text{St}(s) \subseteq \text{St}_2(t)$, then there is an automorphism $\alpha \in \text{Aut}(W)$ such that $\alpha(s) = st$. (Thus (W, S) and $(W, \alpha(S))$ have different sets of reflections.)*
2. *If $\mathcal{V}$ contains a triple of distinct vertices s_1, s_2, s_3 and an edge $[s_2 s_3]$ with label $n > 2$ such that s_2 and s_3 both lie in every maximal spherical simplex containing s_1, then there is an automorphism $\alpha \in \text{Aut}(W)$ such that $\alpha(s_1) = s_1(s_2 s_3)^{n/2}$. (Again, (W, S) and $(W, \alpha(S))$ have different sets of reflections.)*
3. *If no pair or triple of vertices exists as in (1) or (2), respectively, then any two even systems (W, S) and (W, S') yield the same sets of reflections.*

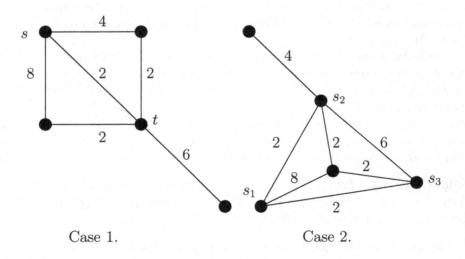

Case 1.   Case 2.

Fig. 3.2   The forbidden configurations of Theorem 3.8

This theorem will allow a complete characterization of reflection preserving even Coxeter systems. In conjunction with Theorem 3.10 (the Even Isomorphism Theorem, due to M. Mihalik), it will also allow a characterization of even reflection independent groups.

The following result (proven in the author's dissertation) is used as a lemma in the proof of the Even Isomorphism Theorem:

Theorem 3.9 *Let W be an arbitrary Coxeter group. Then there is at most one even diagram (up to edge-labeled graph isomorphism) which corresponds to W.*

Theorem 3.9 can actually be strengthened, as we will see in Chapter 5. This theorem can be proven by defining a new combinatorial structure called the pattern of an even Coxeter system. Suppose that (W, S) is an even system with diagram $\mathcal{V}$. Roughly speaking, the *pattern* $\mathcal{P}(\mathcal{V})$ is a collection of edge- and vertex-labeled graphs G_s in one-to-one correspondence with the vertices of $\mathcal{V}$. The graph G_s encodes information about the way in which the simplices containing the vertex s intersect one another. After showing that patterns fully and faithfully record all of the information contained in the diagram (or system), one can show that any two patterns $\mathcal{P}(\mathcal{V})$ and $\mathcal{P}(\mathcal{V}')$ for a group W must be "isomorphic". We will discuss patterns briefly in Chapter 5, but in Chapter 6 we will indicate a more general technique which can be used to prove Theorem 3.9.

Theorem 3.10 **[The Even Isomorphism Theorem]** *Suppose that (W, S) is an even Coxeter system and $\mathcal{V}'$ is a Coxeter diagram for W with an odd edge labeled $[st]$. Then there is a diagram $\mathcal{V}''$ for W obtained from $\mathcal{V}'$ by first (perhaps) performing a diagram twist around the edge $[st]$ and then replacing some triangle $\{[st], [tu], [us]\}$ by an edge with an even label.*

This theorem enables us both to characterize all diagrams which may correspond to a given even Coxeter group W and to describe an algorithm which recovers the unique (by Theorem 3.9) even diagram to which W corresponds. In particular, it allows a characterization of all even rigid Coxeter groups. The following theorem is also easily obtained, as a consequence of the proof of Theorem 3.10:

Theorem 3.11 *Let W be an even Coxeter group. If W is reflection independent, then it is rigid.*

Having characterized even reflection preserving, even reflection independent, and even rigid groups, we move on to the question of strong rigidity.

Theorem 3.4 provides a start; in fact, this theorem hints at a generalization. Regarding more general even groups, we have the following result (from [Bahls (2004a)]). (Here, $Z(G)$ represents the center of the group G, and for $H \leq G$, $C(H)$ denotes the centralizer of the subgroup H relative to the group G, as in Chapter 2.)

Theorem 3.12 *Let (W, S) be an even Coxeter system with connected diagram $\mathcal{V}$. If W is strongly rigid, then W is reflection independent and $\mathcal{V}$ contains no set of vertices J such that the following are both true:*

1. The full subgraph Γ on the vertices $S \setminus J$ has at least 2 connected components, and
2. there are vertices s_1 and s_2 in different connected components of Γ and an element $w \in Z(W_J)$ such that $ws_1 \neq s_1 w$ and $ws_2 \neq s_2 w$.

Moreover, if W is an even reflection independent Coxeter group whose diagram $\mathcal{V}$ has no such sets of vertices, then W is strongly rigid provided $\mathcal{V}$ has at least 3 vertices and satisfies one of the following conditions:

3. $\mathcal{V}$ is of large type, or
4. $\mathcal{V}$ contains no simple circuits of length 3 or 4.

A *circuit of length k* in a graph Γ is any collection C of k distinct edges $\{[s_1 s_2], ..., [s_{k-1} s_k], [s_k s_1]\}$. A circuit C is called *simple* if the each of the vertices s_i ($i = 1, ..., k$) is distinct. Figure 3.3 illustrates the forbidden configuration of Theorem 3.12; here we may take $w = stst$, and s_1 and s_2 as shown.

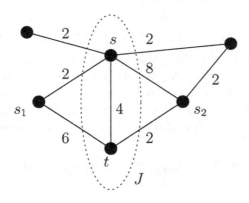

Fig. 3.3   The forbidden configuration of Theorem 3.12

The reader should compare the conditions in the statement of Theorem 3.4 with those in Theorem 3.12. The latter theorem is the first of a number of more general results which can be proven using methods in which the circuits of the diagram $\mathcal{V}$ figure heavily. It is to such results that we now turn.

3.4 Large-type and two-dimensional Coxeter groups

Recall that a Coxeter system (W, S) is said to be of large-type if every edge in the corresponding diagram has label greater than 2. Such systems live at the opposite end of the spectrum from the right-angled groups, but like the right-angled groups they are often much easier to study than arbitrary Coxeter groups. (For instance, the method of small cancellation theory can often be applied to extra-large type groups. See Chapter 6, as well as [Appell and Schupp (1983)], [Kapovich and Schupp (preprint)], [Kapovich and Schupp (2004)], for such applications.)

As we have already noted, every large-type system is two-dimensional. Moreover, some of the methods of proof which apply to large-type systems apply as well to two-dimensional systems.

Regarding large-type groups, we have Theorem 3.13 below. To state this result, we need a little more graph theoretic terminology.

A *spike* in a diagram $\mathcal{V}$ is an edge $[st]$ for which the vertex s has valence 1, so t is its only neighbor in $\mathcal{V}$. We say that a connected diagram $\mathcal{V}$ is *0-connected* if it remains connected upon the removal of any *vertex* of $\mathcal{V}$. We say that a 0-connected diagram $\mathcal{V}$ is *edge-connected* if it remains connected upon the removal of any *edge* of $\mathcal{V}$. (We also say that a 0-connected diagram $\mathcal{V}$ is *odd-edge-connected* if it remains connected upon the removal of any edge *with odd label*. In particular, any edge-connected diagram is also odd-edge-connected.) For instance, see Figure 3.4; the unlabeled edges in this figure can be labeled in any fashion without effecting the connectivity of the given diagram.

The following theorem, whose proof we shall outline in Chapter 6, is due to B. Mühlherr and R. Weidmann.

Theorem 3.13 *Let (W, S) be a large-type Coxeter system with diagram $\mathcal{V}$.*

1. *(W, S) is reflection rigid, up to diagram twisting. (That is, given a system (W, S') which yields the same reflections as (W, S), there is a*

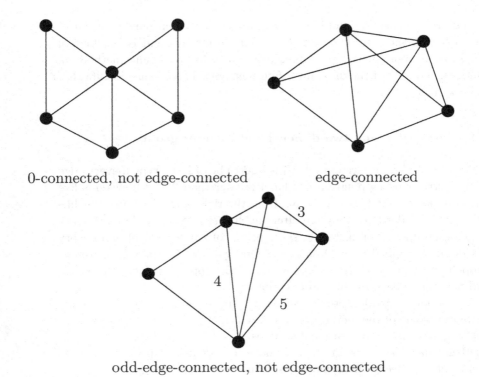

0-connected, not edge-connected   edge-connected

odd-edge-connected, not edge-connected

Fig. 3.4   Various connectivity conditions

sequence of diagram twists which transforms the first system into the second.)

2. *If furthermore* $\mathcal{V}$ *does not have a spike with label* $2(2k+1)$ *for any* $k \geq 1$, *then* W *is rigid up to diagram twisting.*

3. *Let* $|S| \geq 3$. *If* $\mathcal{V}$ *is edge-connected, then* W *is strongly rigid.*

The proof of this theorem makes use of chord-free circuits in the Coxeter diagram of a given group, as well as the visual decomposition of M. Mihalik and S. Tschantz outlined in the previous chapter. In Chapter 6 we will also develop another method (due to the author), called "centralizer chasing", which also depends on chord-free circuits in the diagram $\mathcal{V}$. This method will be used to prove Theorem 3.12 above. These methods, coupled with careful bookkeeping, should allow the proof of yet more general results. For instance, it is conjectured that *all* Coxeter groups are reflection rigid up to diagram twisting, and moreover that the word "reflection" can be removed from this statement if one is to exchange certain edges for triangles (and *vice versa*) in $\mathcal{V}$ in addition to performing twists on $\mathcal{V}$. Very recent work of Howlett and Mühlherr ([Howlett and Mühlherr (2004)]) in fact reduces the Isomorphism Problem to the same problem in the context of reflection preserving systems.

3.5 Related issues

We first turn to the structure of the automorphism group of W (which is clearly intimately related to the rigidity of W). We then examine the implications of our rigidity results for a class of groups related to Coxeter groups, the Artin groups.

3.5.1 *Automorphisms of right-angled Coxeter groups*

As the definitions of Section 3.1 clearly demonstrate, questions concerning rigidity and strong rigidity of a group W are in fact questions concerning the structure of $\mathrm{Aut}(W)$. To ask if W is rigid is to ask if W has sufficiently many automorphisms to describe all possible Coxeter systems for W; to ask if W is strongly rigid is to ask if W has a sufficient supply of inner automorphisms for such a purpose. Questions about $\mathrm{Aut}(W)$ are therefore in some sense refinements of questions about rigidity: knowing $\mathrm{Aut}(W)$, we can know whether or not W is rigid or strongly rigid.

In this subsection we consider some of the results concerning $\mathrm{Aut}(W)$ for various classes of Coxeter groups. These results will be proven in Chapter 7. Once more we begin our discussion with the right-angled groups.

One of the first results proven concerning the structure of $\mathrm{Aut}(W)$ for a Coxeter group W was Theorem 3.14 (due to J. Tits), which allows $\mathrm{Aut}(W)$ to be decomposed as a semidirect product involving simpler groups. In order to explain this theorem, we need some new notation.

As in 1.3.3, denote by $\mathcal{S} = \mathcal{S}(W)$ the collection of spherical subsets of $\mathcal{V}$ (which in the right angled case coincides with the set of simplices of $\mathcal{V}$). We may define a partial product, $\cdot$, on $\mathcal{S}$ by $S_1 \cdot S_2 = (S_1 \setminus S_2) \cup (S_2 \setminus S_1)$ for $S_i \in \mathcal{S}$ (this is no more than the symmetric difference of S_1 and S_2). As we will see in Chapter 7, $(\mathcal{S}, \cdot)$ is a finite groupoid, and therefore its automorphism group $\mathrm{Aut}(\mathcal{S})$ is also finite.

Remark 3.1 The term *groupoid* has multiple meanings in the literature;

see Section 7.1.1 for the use of the term in our work.

We define the subgroup $\text{Spe}(W)$ of $\text{Aut}(W)$ to be the kernel of the natural action of $\text{Aut}(W)$ on the set of conjugacy classes of involutions. The elements of $\text{Spe}(W)$ are called the *special automorphisms* of W.

Theorem 3.14 *Let W be a right angled Coxeter group with diagram $\mathcal{V}$. Then* $\text{Aut}(W)$ *is a semidirect product of* $\text{Aut}(\mathcal{S})$ *by* $\text{Spe}(W)$.

Theorem 3.14 is in fact a generalization of a similar result proven by L. D. James in her dissertation, [James (1985)]. In Chapter 7 we will go beyond the above theorem and develop an explicit presentation for $\text{Spe}(W)$ whenever W is right angled (given by Theorem 7.2), so that $\text{Aut}(W)$ can be completely understood in this case.

3.5.2 *Automorphisms of more general Coxeter groups*

What of $\text{Aut}(W)$ in general? For broader classes of groups, $\text{Aut}(W)$ proves much more difficult to understand. There are a few positive results, however. In [Howlett, *et al.* (1997)], the following theorem is proven:

Theorem 3.15 *Let (W, S) be a Coxeter system in which every product st $(s, t \in S)$ has finite order. Then* $|\text{Out}(W)|$ *is finite.*

Theorem 3.15 essentially reduces study of $\text{Aut}(W)$ to study of $\text{Inn}(W)$, which is often easy to describe (since the center $Z(W)$ is nearly always trivial).

W. Franzsen ([Franzsen (2001)]) and Franszen and Howlett ([Franzsen and Howlett (2001)], [Franzsen and Howlett (2003)]) have proven a number of other results concerning automorphisms of Coxeter groups which are "small" in some way, either by restricting the size of the fundamental generating set, or by assuming that the Coxeter group contains a spherical subgroup of rank nearly that of the group itself.

We have already mentioned (Corollary 3.1) the fact that $\text{Aut}(W)$ is very simple in case W is strongly rigid. This simple structure stems from the fact that strong rigidity forces any automorphism to act almost like conjugation. In case W is not strongly rigid but still has a somewhat "rigid" structure, a typical automorphism may not be inner, but may behave like an inner automorphism on each of several different portions of the group which are themselves strongly rigid.

This is the idea behind the following theorem from [Bahls (2004b)]. A Coxeter diagram $\mathcal{V}$ is said to have *no vertex branching* (or have NVB) if the removal of any vertex of $\mathcal{V}$ (along with any incident edges) separates $\mathcal{V}$ into no more than 2 components. (Any vertex whose removal results in more than a single component is called a *cut vertex*.)

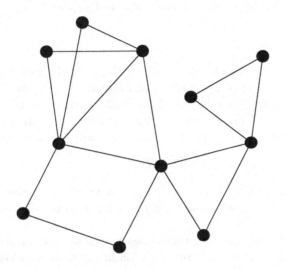

Fig. 3.5   A diagram satisfying NVB

Theorem 3.16   *Let W be an even, large-type Coxeter group whose diagram is connected and has NVB. Then* $\mathrm{Aut}(W)$ *is a semidirect product of G with* $\mathrm{Diag}(W)$, *where G is, up to a subgroup of finite index, the product of* $\mathrm{Inn}(W)$ *and certain subgroups of centralizers of edges and vertices of* $\mathcal{V}$.

More precisely, this theorem allows a characterization of the even, large type, NVB groups for which $\mathrm{Out}(W)$ is finite:

Corollary 3.2   *Let W be as in Theorem 3.16. Then the following statements are true.*

1. $\mathrm{Out}(W)$ *contains a subgroup of finite index isomorphic to a direct product of free powers of the two-element group* $\mathbb{Z}_2$.
2. $\mathrm{Out}(W)$ *is infinite if and only if there is a cut vertex s in* $\mathcal{V}$ *such that s is adjacent to at least two other vertices in* $\mathcal{V}$.
3. *Suppose further that* $\mathcal{V}$ *has no cut vertices. then* $\mathrm{Aut}(W)$ *is isomorphic to a semidirect product of* $\mathrm{Diag}(W)$ *by* $\mathrm{Inn}(W) \times (\mathbb{Z}_2)^k$, *for some integer k.*

Theorem 3.16 is closely related to work of G. Levitt on the automorphism groups of various classes of groups, including one-ended word hyperbolic groups and generalized Baumslag-Solitar groups (see [Levitt (preprint)]). We will address these connections in Chapter 7.

3.5.3   *Artin groups*

The class of Artin groups is closely related to the class of Coxeter groups. Let $S = \{s_i\}_{i \in I}$ be a set of generators, and assign to each pair s_i, s_j of distinct generators from S a value m_{ij} from $\{2, 3, ..., \infty\}$ (note that $m_{ij} = m_{ji}$). Given $s_i, s_j \in S$, define the word $u(i, j)$ to be the alternating product $s_i s_j s_i \cdots$ in which exactly m_{ij} letters appear. We now define a group

$$A = \langle S \mid R \rangle \tag{3.3}$$

where $R = \{u(i, j) = u(j, i) \mid i, j \in I; i \neq j\}$. A group possessing such a presentation is called an *Artin group*, and the pair (A, S) is called an *Artin system*.

Clearly Artin groups bear structural resemblance to Coxeter groups; if we add the relations $s_i^2 = 1$ to the presentation in (3.3), the resulting quotient group is easily seen to be a Coxeter group, called the Coxeter group *associated to A*. The resemblance between these groups allows us to apply much of the terminology and machinery used to understand Coxeter groups to Artin groups instead. In particular, each Artin system (A, S) is associated with an edge-labeled diagram which fully and faithfully represents the presentation corresponding to (A, S), and the notions of evenness, reflections, (reflection) rigidity, strong (reflection) rigidity, and diagram twisting all still make sense. (Even the Davis complex, suitably modified, can be defined, although it is no longer an easy matter to impose a nice metric structure upon it.) We say an Artin group A is *of finite-type* if the Coxeter group associated to it is finite.

Despite their similarities, Coxeter groups and Artin groups are very different creatures. Far less is known of the structure of Artin groups. For instance, it is not even known yet (however likely the prospect) whether the Word Problem is solvable for every Artin group. Nor is it known whether (as is also conjectured) every Artin group is torsion free. There are, however, some "Artin analogues" to the results that we have considered in this chapter.

One such analogous result is the following, due to N. Brady, *et al.* (see [Brady, *et al.* (2002)]). By *diagram* we mean any edge-labeled graph that could be the Coxeter (Artin) diagram for some Coxeter (Artin) system (W, S) $((A, S))$.

Theorem 3.17 *Let $\mathcal{V}$ be a diagram; let (W, S) be the corresponding Coxeter system, and let (A, S) be the corresponding Artin system. If (W, S) is reflection rigid up to diagram twisting, then (A, S) is reflection rigid up to diagram twisting.*

In conjunction with Theorem 3.5, we obtain the following result.

Theorem 3.18 *Let $\mathcal{V}$ be the diagram corresponding to the Artin system (A, S). If either of the following statements is true, then (A, S) is reflection rigid:*

1. A is of finite type, or
2. (A, S) is even.

If instead $\mathcal{V}$ is a tree, then (A, S) is reflection rigid, up to diagram twisting.

We obtain similar results for Artin groups by applying Theorem 3.17 to Theorem 3.13.

Recently, L. Paris has obtained a result which gives an analogue of Radcliffe's proof that finite irreducible Coxeter groups are isomorphic if and only if their diagrams are isomorphic as graphs. We will consider this result briefly in Chapter 7.

It is unclear how many of the methods used to investigate the rigidity of Coxeter groups can be applied successfully to Artin groups. Such questions lie at the frontier of current research in this field.

3.6 Exercises

1. Prove that W is rigid if and only if its diagram, $\mathcal{V}$, is unique up to edge-labeled graph isomorphism.

2. Use Theorem 3.12 to prove that the group with presentation (3.2) is not strongly rigid.

3. Prove that an infinite group of isometries of a CAT(0) space X need not fix some point of X.

4. Give an example of a Coxeter system (W, S) and subset $T \subseteq S$ such that there is no retraction from W onto W_T. (*Hint*: consider a triangle group.)

5. Prove the first two statements made in Theorem 3.8.

6. Let S' be the generating set obtained from the fundamental generating set S (for system (W, S)) by the diagram twist described in Section 3.3. Show that (W, S') is also a Coxeter system by showing that it corresponds to the diagram $\mathcal{V}'$ described in that section.

7. Show that if (W, S) is even, then it admits no nontrivial diagram twists.

8. Prove that for any group G, $\mathrm{Inn}(G) \leq \mathrm{Spe}(G) \leq \mathrm{Aut}(G)$, where $\mathrm{Spe}(G)$ is defined as in 3.5.1.

Chapter 4

In the beginning: some early results

In this chapter we begin to prove the theorems presented in Chapter 3. The proofs showcase a number of techniques, from straightforward combinatorics (Theorems 4.1, 4.2, and 4.3) to more sophisticated homological arguments (Theorem 4.4).

Many of the early theorems rely upon intermediate results (such as Lemma 1.2 and Proposition 4.1) which will later be generalized in order to deal with broader classes of groups.

Unless otherwise noted, the diagram corresponding to the Coxeter system (W, S) will be denoted by $\mathcal{V}$. Moreover, we will consistently make use of "primed" and "unprimed" notation, so that the system (W, S') will correspond to the diagram $\mathcal{V}'$, and so forth.

4.1 Rigidity of right-angled Coxeter groups

We begin by considering the simplest of all (infinite) Coxeter groups, the right-angled groups. As shown below (Theorem 4.1), the Isomorphism Problem for right-angled Coxeter groups is trivial. Strengthening this with a characterization of strongly rigid right-angled Coxeter groups is a more difficult problem, to which we will return later (see Section 5.4).

In his dissertation ([Radcliffe (2000)]), D. Radcliffe proved the following result concerning right-angled Coxeter groups:

Theorem 4.1 *Let W be a right-angled Coxeter group. Then W is rigid.*

Using Lemma 1.2 it is not difficult to establish the generalization of Proposition 3.1 promised in 3.2.2. Given a Coxeter system (W, S), we denote by $M(W, S)$ the set of maximal spherical simplices in the diagram $\mathcal{V}$. (That is, $\sigma \in M(W, S)$ if and only if W_σ is spherical and if τ is any simplex properly containing σ, then W_τ is infinite.)

Proposition 4.1   *Let (W, S) and (W, S') be systems for an arbitrary Coxeter group W. Then there is a one-to-one correspondence $\phi : M(W, S) \to M(W, S')$ such that for every $\sigma \in M(W, S)$ there is a group element $w \in W$ for which $W_\sigma = wW_{\phi(\sigma)}w^{-1}$.*

Proof.   Given $\sigma \in M(W, S)$, apply Lemma 1.2 to obtain $w' \in W$ and $\sigma' \subseteq \mathcal{V}'$ such that $W_\sigma \subseteq w'W_{\sigma'}w'^{-1}$. Apply the lemma again to obtain $w \in W$ and $\sigma'' \subseteq \mathcal{V}$ such that $W_{\sigma'} \subseteq wW_{\sigma''}w^{-1}$. Thus $W_\sigma \subseteq w'wW_{\sigma''}(w'w)^{-1}$. It is now easy to show that the maximality of σ forces equality to hold in this relation, so that $W_\sigma = w'W_{\sigma'}w'^{-1}$. Moreover, it is clear that $\sigma' \in M(W, S')$. By using the fact that the action of W on $\Sigma(W, S)$ is properly discontinuous, one can show that no two distinct maximal spherical subgroups (with respect to a given system) are conjugate to one another. The map ϕ defined by $\phi(\sigma) = \sigma'$ then provides the desired bijection.   □

We will make frequent use of this proposition throughout the text, often without explicit mention.

Now let (W, S) be a right-angled Coxeter system. Let $\Sigma = \Sigma(W, S)$ and let (W, S') be any other system for W. Using Lemma 1.2 it is easy to show that (W, S') is also right-angled and that $|S| = |S'|$.

Following Radcliffe, we now consider the natural quotient map $\nu : W \longrightarrow W/W'$, where W' is the commutator subgroup of W. We may identify W/W' with $(\mathbb{Z}_2)^{|S|}$. It is obvious that ν is injective on each spherical subgroup of W. Moreover, if ϕ is the map whose existence is guaranteed by Proposition 4.1, then

$$\nu(W_\sigma) = \nu(W_{\phi(\sigma)}) \tag{4.1}$$

for every $\sigma \in M(W, S)$. From this, it is easy to establish the following result.

Proposition 4.2

1. Let $\sigma_1, ..., \sigma_k \in M(W, S)$. Then

$$\left| \bigcap_{i=1}^{k} \sigma_i \right| = \left| \bigcap_{i=1}^{k} \phi(\sigma_i) \right|.$$

2. *Let* $\sigma_1, ..., \sigma_k, \tau_1, ..., \tau_l \in M(W, S)$. *Then*

$$\left| \bigcap_{i=1}^{k} \sigma_i \setminus \bigcup_{j=1}^{l} \tau_j \right| = \left| \bigcap_{i=1}^{k} \phi(\sigma_i) \setminus \bigcup_{j=1}^{l} \phi(\tau_j) \right|.$$

Proof. The first statement is obtained by applying Proposition 1.4 and using the fact that ν is one-to-one on spherical subgroups. The second statement follows from the first by an inclusion-exclusion argument and induction on l. The details are left to the reader. □

Roughly speaking, Proposition 4.2 tells us that "spherical subgroups of intersections of conjugate subgroups are conjugate". The proof of Theorem 4.1 is now almost immediate. Indeed, given $s \in S$, Proposition 4.2 implies that there is a unique generator $s' \in S'$ such that

$$s \in \sigma \Leftrightarrow s' \in \phi(\sigma) \tag{4.2}$$

for every $\sigma \in M(W, S)$. Given $s, t \in S$, it is easy to show that st and $s't'$ have the same order (either 2 or ∞), so that the bijection given in (4.2) extends to an automorphism of W mapping S to S'. Theorem 4.1 is now proven.

4.2 A slight generalization

Using similar methods, Radcliffe was able to prove the following generalization of his initial result:

Theorem 4.2 *Let (W, S) be a Coxeter system such that every edge in the corresponding diagram $\mathcal{V}$ is labeled by either 2 or by an integer divisible by 4. Then W is rigid.*

This theorem is related to another, for which the hypotheses are decidedly different:

Theorem 4.3 *Let (W, S) be a Coxeter system such that the label of any edge in the corresponding diagram $\mathcal{V}$ is either odd or divisible by 4. Then W is reflection independent.*

The proof of this theorem is elementary. We will outline the proof, leaving the details to the exercises. The theorem will follow from a trio of easy lemmas.

Lemma 4.1 *Let $W_{\{s,t\}} = wW_{\{s',t'\}}w^{-1} \cong D_k$, $k \geq 3$. Then each generator in $\{s,t\}$ is conjugate to some generator in $\{s',t'\}$.*

Proof. We merely need to examine the elements of order two in D_k. Every such element is a reflection in case k is odd. If k is even, $(s't')^{k/2}$ has order two, but it is central in $W_{\{s',t'\}}$, therefore cannot be conjugate to either s or t.   □

Lemma 4.2 *Let $\sigma \subseteq V$ be a spherical simplex. Then either σ contains a triangle with multiset of edge labels $T_1 = \{2,3,3\}$, $T_2 = \{2,3,4\}$, or $T_3 = \{2,3,5\}$, or W_σ is isomorphic to a direct product of dihedral groups and the two-element group $\mathbb{Z}_2$.*

Proof. Exercise. (Recall Lemma 1.1.)   □

Lemma 4.3 *Let $n \geq 3$.*

1. *If D_n decomposes nontrivially as a direct product $H \times K$, then $n = 2k$ for some odd integer k, $H \cong D_k$, and $K \cong \mathbb{Z}_2$.*
2. *If $W_\sigma \cong D_n$ does not decompose nontrivially as a direct product, then σ must be an edge labeled n.*

Notice that the first statement justifies our exclusion of edges with labels $2k$, where k is odd.

Proof. We leave the proof of the first statement as an exercise (*cf.* Exercise 1.18). As for the second, because of Lemma 4.2 we need only show that no triangle with edge labels T_i appears in σ. This can be done in any number of ways (by considering derived lengths or commutator subgroups, presentations, *et cetera*).   □

We can now finish the proof of Theorem 4.3.

Consider a second diagram, V', corresponding to W. A simple quotient argument shows that we may assume neither V nor V' has any single-vertex components (another exercise!). Let σ' be any maximal spherical simplex in V'. By Lemma 1.1 and Proposition 4.1, there is a unique edge $\sigma \subseteq V$ such that $W_{\sigma'} = wW_\sigma w^{-1}$ for some $w \in W$. Thus $W_{\sigma'} \cong D_n$ does not decompose nontrivially as a direct product, by the first statement of Lemma 4.3, and by the second statement of the same result, σ' must be an edge with the same label as σ. Applying Lemma 4.1 shows that each reflection in $W_{\sigma'}$ is a reflection in W, as desired.

4.3   Strong rigidity through homological conditions

Now for a change of course.

In this section we will outline a proof of the following theorem from [Charney and Davis (2000)]:

Theorem 4.4   *Suppose that a Coxeter group is of type* HM_n *or* PM_n *(as defined in the following subsection). Then* W *is strongly rigid.*

The proof we present underscores the strong connection between Coxeter groups and the topology of manifolds. Indeed, the results of whose proofs are sketched below follow closely the earlier work of M. Davis (including [Davis (1983)] and [Davis (1998)]) regarding the applications of Coxeter groups to low-dimensional topology.

4.3.1   *The conditions* HM_n *and* PM_n

We first require a careful definition of a "combinatorial neighborhood". Given an abstract simplicial complex C and a k-simplex σ in C, the *link* of the simplex σ in C (denoted $\mathrm{Lk}(\sigma, C)$) is the abstract simplicial complex whose simplices are in one-to-one correspondence with the simplices of C which properly contain σ. It is left to the reader to check that this definition makes sense, and to investigate the dimension of the resulting complex.

A space X is called a *homology n-manifold* if for every point $x \in X$, the relative integral homology group $H_i(X, X \setminus \{x\})$ is $\mathbb{Z}$ if $i = n$, and 0 otherwise. In case X is a simplicial complex of dimension n, this is equivalent to saying that for every k-simplex σ in X, $\mathrm{Lk}(\sigma, X)$ is homologically indistinguishable from the sphere $\mathbb{S}^{n-k-1}$. If the homology n-manifold X itself has the same homology as the sphere $\mathbb{S}^n$, we call X a *generalized homology n-sphere*.

Now let (W, S) be a Coxeter system. We say that (W, S) is of type HM_n if the nerve $N(W, S)$ is a generalized homology $(n-1)$-sphere. (The nerve is isomorphic to the link of a vertex of the Davis complex; see Exercise 1.13. Thus, the nerve having dimension $n-1$ is equivalent to the complex $\Sigma(W, S)$ itself having dimension n, justifying the index n in the notation HM_n.)

For example, if the diagram $\mathcal{V}$ consists of a single simple circuit $\{s_1, ..., s_k\}$, where $k \geq 4$, the nerve $N(W, S)$ is isomorphic to $\mathcal{V}$ (as an unlabeled graph); hence, $N(W, S)$ is homeomorphic to a 1-sphere, and the link of each vertex in $N(W, S)$ is homeomorphic to a 0-sphere, while the link of each edge is empty. Thus $N(W, S)$ is a homology 1-sphere, so that

W is of type HM_2.

Consider instead a Coxeter system whose nerve consists of a triangulated 2-sphere, perhaps as in Figure 4.1, where the nerve $N(W, S)$ is an icosahedron. (This figure obtains, for instance, when W is right-angled and $\mathcal{V}$ is the 1-skeleton of the icosahedron.)

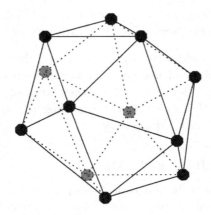

Fig. 4.1   The nerve of a Coxeter system of type HM_3

It is easy to see that W is of type HM_3 in this case.

In particular, then, Theorem 4.4 will prove the strong rigidity of these classes of Coxeter groups W. In [Charney and Davis (2000)] it is shown that W is of type HM_n if and only if W acts effectively, properly, and cocompactly on a contractible manifold. (See the proof of Proposition 4.4, below.)

Consider now a *locally finite* simplicial complex X. That is, each simplex of X is contained in only finitely many other simplices. We will call X a *pseudo-n-manifold* if every maximal simplex of X has dimension n and every $(n-1)$-simplex is a face of exactly two n-simplices. (Obviously such properties are true of a triangulated n-manifold without boundary.) If it is possible to choose orientations on the n-simplices in X such that the sum of all such simplices is a (possibly infinite) n-cycle, we call X *orientable*, and an *orientation on* X is any choice of orientation on the individual simplices.

Let (W, S) be a Coxeter system. We say that (W, S) is of type PM_n if the nerve $N(W, S)$ is an orientable pseudo-n-manifold and $H_{n-1}(N(W, S)) \cong \mathbb{Z}$. It is left to the reader to check that if (W, S) is of type HM_n, then it is of type PM_n.

Because the conditions HM_n and PM_n force the Davis complex $\Sigma(W, S)$ to have a rather simple structure, it is not difficult to obtain a great deal of geometric intuition regarding the proof of Theorem 4.4. In the remainder of this section we will outline the major steps in the argument. We will prove only the case in which W is of type HM_n, referring the reader to [Charney and Davis (2000)] for the proof in case W is of type PM_n.

4.3.2 *Fixed point sets and the nerve of (W, S)*

The idea of the proof is the following: assume we are given the systems (W, S) and (W, S'). We first show that under the hypothesis of the theorem, both systems yield the same reflections (*i.e.*, that W is reflection independent.) Therefore the generators S' act as reflections in the walls of $\Sigma(W, S)$ as well as in the walls of $\Sigma(W, S')$. We then show that this action has as its fundamental domain a translate of the fundamental domain of the action of (W, S) on $\Sigma(W, S)$, from which it follows that S and S' are conjugate to one another.

Before we consider Coxeter groups of type HM_n and PM_n, we state a couple of results concerning the structure of $\Sigma(W, S)$ for a general system (W, S).

Fix the system (W, S). For any given spherical subset $T \subseteq S$, define

$$F(T) = \{wW_{T'} \in W\mathcal{S} \mid W_T w W_{T'} = w W_{T'}\}. \tag{4.3}$$

Thus $F(T)$ is nothing other than the poset of spherical cosets (that is, vertices of Σ) which are left invariant by the action of W_T.

Lemma 4.4 *Let $T \in \mathcal{S}$ and let $wW_{T'} \in F(T)$ as above. Then $\downarrow_{F(T)} wW_{T'}$ is isomorphic to the poset of faces of a convex cell of dimension $|T'| - |T|$.*

Proof. We make use of the Coxeter cells defined in 1.5.4. The easiest case to visualize is the case in which $T = \emptyset$; in this case $F(T) = W\mathcal{S}$ and $\downarrow_{F(T)} wW_{T'}$ is isomorphic to the poset of faces of the Coxeter cell corresponding to $W_{T'}$. In case T is not empty, $\downarrow_{F(T)} wW_{T'}$ will be isomorphic to a subposet of the poset of faces of the Coxeter cell corresponding to $W_{T'}$. (Namely, to the subposet consisting of those faces which intersect the subspace fixed by $w^{-1}W_T w$.) $\qquad\square$

To illustrate, consider the Coxeter group $(3, 3, 3)$ with the presentation given in 1.5.4. Here, Σ is homeomorphic to the Euclidean plane. When T

consists of a single generator $x \in \{a, b, c\}$, the fixed point set Σ^x is a line. Let $wW_{T'}$ be some coset lying on this line. If T' also consists of a single generator, then $\downarrow_{F(T)} wW_{T'}$ is a single vertex (lying on an edge of a maximal Coxeter cell in Σ). If T' consists of a pair of generators, $\downarrow_{F(T)} wW_{T'}$ is a three-element poset isomorphic to the poset of faces of a single edge. This is shown in Figure 4.2, where $T = \{a\}$ and $wW_{T'} = W_{\{a,b\}}$. If T is empty, of course $F(T) = WS$, and $\downarrow_{F(T)} wW_{T'}$ is precisely the poset of faces of the hexagonal Coxeter cell corresponding to $wW_{T'}$.

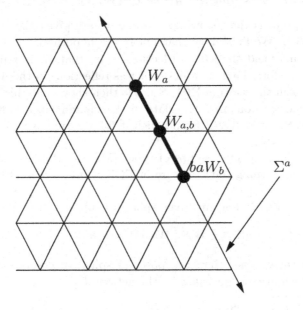

Fig. 4.2   Lemma 4.4: the heavily shaded segment represents $\downarrow_{F(T)} wW_{T'}$

As a consequence of Lemma 4.4, we may define a cell structure on the fixed point set Σ^{W_T} of the action of W_T on Σ, in the natural fashion. In fact, the links of vertices in the complex so defined resemble the links of various simplices in $N(W, S)$ (*cf.* Exercise 1.13):

Lemma 4.5   *If $T \in S$, the cell structure on Σ^{W_T} defined by means of Lemma 4.4 satisfies*

$$\mathrm{Lk}(wW_{T'}, \Sigma^{W_T}) \cong \mathrm{Lk}(\sigma_{T'}, N(W, S))$$

for every vertex $wW_{T'}$ in Σ^{W_T}. (Here, $\sigma_{T'}$ is the simplex of the nerve *corresponding to T'.)*

Lemma 4.5 allows us to pass structure of the nerve onto structure of the fixed point sets of spherical subgroups. In particular, taking $T = \emptyset$, the link of any given simplex in $\Sigma = \Sigma^{W_T}$ is isomorphic to the link of the appropriate simplex in $N(W, S)$.

4.3.3 *The action of (W, S') on Σ*

Now assume that W is of type HM_n. Because $N(W, S)$ is a simplicial complex in which the links of simplices are nice, the links of simplices of Σ are very well-behaved, and we obtain the following result.

Proposition 4.3 *Let (W, S) be of type HM_n. Then Σ is a homology n-manifold. (Moreover, each fixed point set Σ^{W_T} is a homology $(n - k)$-manifold, if T has rank k.)*

It is at this point that a we begin to draw upon the homology theory discussed briefly in Chapter 2. We show that both the set of reflections and the property of being of type HM_n are independent of the choice of system.

Proposition 4.4 *Suppose that (W, S) is of type HM_n, and let (W, S') be any other system for W. Then the following are true.*

1. *(W, S') is of type HM_n, and*
2. *(W, S) and (W, S') yield the same sets of reflections.*

Let us at least sketch a proof.

Proof. It can be shown (see Theorem 15.3 of [Davis (1983)] and the introduction of [Davis (1998)]) that (W, S) is of type HM_n if and only if W acts effectively, properly, and cocompactly on some contractible manifold. The latter condition is independent of the presentation, so that the first statement in Proposition 4.4 follows.

The second statement follows from computations involving the virtual cohomological dimension of certain subgroups of W:

Lemma 4.6 *Given any Coxeter group W of type HM_n, the spherical parabolic subgroups of W of rank k are precisely the maximal finite subgroups G of W with the property that*

$$\text{vcd}(N(G)) = n - k,$$

where $N(G)$ is the normalizer of G in W.

From this it follows that the rank-1 spherical parabolic subgroups (namely, the reflections) of W are independent of the choice of system.

Proof. Let G be a spherical parabolic subgroup of rank k. As mentioned in 2.1.1, Σ^G is contractible. In fact, in our case, Proposition 4.3 says that it is a homology manifold of dimension $n - k$. This is enough to allow us to apply Poincaré duality (see 2.1.2) and conclude that $H_c^i(\Sigma^G)$ is trivial unless $i = n - k$, and $H_c^{n-k}(\Sigma^G) = \mathbb{Z}$.

Now consider the group $N(G)$. This group acts cocompactly on Σ^G with finite isotropy groups (prove this!). From this it follows that Σ^G is the universal cover of a finite $K(H, 1)$-complex, for any subgroup H of finite index in $N(G)$ (*cf.* the example following Theorem 2.1). Much as in 2.1.1, it can be shown that

$$H^i(H; \mathbb{Z}H) \cong H_c^i(\Sigma^G)$$

for all i, proving that $\mathrm{vcd}(N(G)) = n - k$.

Now let $\mathrm{vcd}(N(G)) = n - k$, and let G be maximal among finite subgroups of W with this property. By the fact that $G \leq wW_T w^{-1}$ for some spherical subset $T \subseteq S$ (and the proof just given), $G = wW_T w^{-1}$ must hold. □

Proposition 4.4 now follows. Moreover, the second statement in the proposition can be strengthened: both (W, S) and (W, S') admit the same spherical parabolic subgroups of a given rank k, $k = 1, ..., n$. □

Now we consider two systems (W, S) and (W, S') for the same group of type HM_n, with the aim of showing that S and S' are conjugate. By Proposition 4.4, each generator $s' \in S'$ acts on $\Sigma = \Sigma(W, S)$ as a reflection in some wall $\Sigma^{s'}$, $s' = wsw^{-1}$ ($w \in W, s \in S$). Each such wall $\Sigma^{s'}$ separates Σ into two half-spaces, which we may denote by $H_{s'}^-$ and $H_{s'}^+$. Theorem 4.4 will follow from the following fact.

Theorem 4.5 *There exists a unique function* $\epsilon : S' \to \{+, -\}$ *such that the set*

$$D = \bigcap_{s' \in S'} H_{s'}^{\epsilon(s')}$$

is nonempty, and consists of a single chamber of Σ.

That is, some W-translate of the fundamental chamber K of Σ is a fundamental chamber for the action of (W, S') on Σ.

In order to prove Theorem 4.5, it is useful to make use of galleries (*cf.* 1.5.2). Let X be an n-dimensional simplicial complex which is either a homology manifold or a pseudo-manifold. A *gallery* in X is a sequence $\{\sigma_1, ..., \sigma_k\}$ of n-simplices of X such that σ_i and σ_{i+1} are adjacent for $i = 1, ..., k-1$, simplices σ and τ being called *adjacent* if σ and τ share a common codimension-1 face. (Thus, galleries are essentially combinatorial paths of simplices of maximal dimension.) Given a vertex x in X, we say that two n-simplices σ and τ are *v-connected* if there is a gallery from σ to τ which never intersects the open star of v in X.

We are interested primarily in galleries in the nerve $N(W, S')$ and in Σ. Specifically, by examining galleries in $N(W, S')$, we can determine when two points of Σ lie on the same side of the wall $\Sigma^{s'}$ of Σ corresponding to a generator $s' \in S'$ (Lemma 4.7). Consider any conjugate wW_Tw^{-1} of a maximal spherical subgroup in (W, S) (*i.e.*, a maximal element in $W\mathcal{S}$). By the definition of Σ, it is clear that the fixed point set $\Sigma^{wW_Tw^{-1}}$ of the action of wW_Tw^{-1} on Σ consists of a single point (namely, the intersection of the hyperplanes $w\Sigma^t$ for $t \in T$).

Lemma 4.7 *Suppose that W_{T_1} and W_{T_2} are maximal spherical subgroups in (W, S') (and are therefore conjugate to maximal spherical subgroups in (W, S)). Suppose also that $s' \in S' \setminus (T_1 \cup T_2)$. If the simplices σ_1 and σ_2 in $N(W, S')$ corresponding to T_1 and T_2 are v-connected (where v is the vertex of $N(W, S')$ corresponding to s'), then the points $\Sigma^{W_{T_1}}$ and $\Sigma^{W_{T_2}}$ in Σ lie on the same side of the hyperplane $\Sigma^{s'}$.*

Proof. Inducting on the length of the gallery between σ_1 and σ_2, we can assume that these simplices are adjacent in $N(W, S')$, so that $W_{T_1 \cap T_2}$ is a spherical subgroup of (W, S') of rank $n - 1$. (Recall that we are assuming W to be of type HM_n!) From the proof of Proposition 4.4, $W_{T_1 \cap T_2}$ is a parabolic subgroup of (W, S) of rank $n - 1$, so that $\Sigma^{W_{T_1 \cap T_2}}$ is a 1-dimensional subspace, Y, of Σ. Can Y and $\Sigma^{s'}$ intersect? If this were so, $(T_1 \cap T_2) \cup \{s'\}$ would generate a finite subgroup of W. (Why?) However, this cannot be, by the definition of the nerve $N(W, S')$ and the fact that W is of type HM_n. □

The reader is encouraged to play a simple example (such as the affine group $(3, 3, 3)$) in order to develop some intuition regarding this lemma.

If we remove the star of any vertex v from a generalized homology n-sphere X, the resulting space has the same homology as an $(n-1)$-manifold with boundary and is acyclic. Therefore, $X \setminus \text{Star}(v)$ is connected, and any two $(n-1)$-dimensional simplices can be connected by a gallery. Therefore we can apply Lemma 4.7 to *any* maximal spherical subgroups of (W, S') which do not contain a given generator s' (corresponding to the vertex v whose star is removed from $N(W, S')$).

Fix $s' \in S'$, and let $T \subseteq S'$ be any maximal spherical subset of S'. If $s' \in T$, then Σ^{W_T} lies on the hyperplane $\Sigma^{s'}$, and so lies in either closed half-space $H_{s'}^{\pm}$ into which $\Sigma_{s'}$ separates Σ. Furthermore, Lemma 4.7 implies that every Σ^{W_T} which does not lie on the hyperplane $\Sigma^{s'}$ lies on the same side of $\Sigma^{s'}$. Therefore we can choose *one* of the closed half-spaces $H_{s'}^{\pm}$ which contains *every* Σ^{W_T}, $T \subseteq S'$ a maximal spherical subset. Let $\epsilon(s')$ denote the appropriate choice of sign, and define

$$D = \bigcap_{s' \in S'} H_{s'}^{\epsilon(s')}.$$

Proposition 4.5 D *is a finite union of chambers of* Σ.

The scenario as we know it so far is illustrated in Figure 4.3.

Proof.

Clearly D is nonempty, for otherwise W is finite.

First we show that no proper subspace of Σ contains D (in other words, that D has the appropriate dimension). Suppose this is not the case. Then D lies in the fixed point set $\Sigma^{w W_T w^{-1}}$ of some parabolic subgroup $w W_T w^{-1} \neq W_\emptyset = \{1\}$, by the definition of Σ. For each $T \subseteq S'$, W_T a maximal parabolic subgroup of (W, S'), the fixed point set Σ^{W_T} lies in D, by construction. Let D' be the smallest subspace of Σ containing each such Σ^{W_T}. By the assumption above, D' is contained in a proper subspace of Σ. But it is not difficult to see that D' is $\Sigma^{W_{\overline{T}}}$, where $\overline{T} = \cap T$, the intersection taken over all $T \subseteq S'$ which generate maximal parabolic subgroups. Our assumption implies that $\overline{T} \neq \emptyset$, so every $n-1$ simplex of the nerve $N(W, S')$ contains the non-trivial simplex $\sigma_{\overline{T}}$ of $N(W, S')$. This cannot be, because $N(W, S')$ is a generalized homology sphere. Thus D' is Σ after all.

Why must D be a *finite* union of chambers? First, we define the boundary ∂D of D in a natural way. Namely, we let ∂D be the union (taken over $s' \in S'$) of the sets $D \cap \Sigma^{s'}$. (This should remind the reader of the boundary of K itself, consisting of the mirrors K_s.) In a similar fashion we may define the "faces" of D, yielding a simplicial structure on D comparable to that of Σ. We can then show that $(D, \partial D)$ is a contractible homology n-manifold with boundary, and moreover that ∂D has the same homology as the boundary of the fundamental chamber K' in Σ'. Using once more the fact that W is of type HM_n (and that this is independent of the system

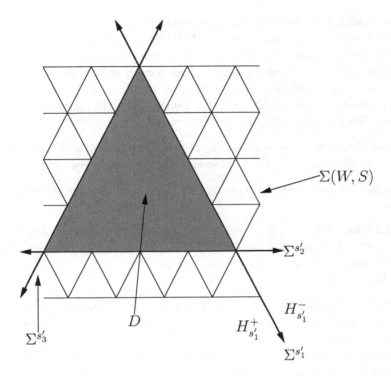

Fig. 4.3   The fundamental domain D

(W, S), as in Proposition 4.4), we conclude that $H_n(D, \partial D) \cong \mathbb{Z}$, so that D is compact, and therefore can contain only finitely many chambers. $\square$

4.3.4   *Completing the proof*

Now, exactly how many chambers *does* D contain? In order to answer this question, we make use of the construction of Tits and Vinberg alluded to in Section 1.5.3.

Suppose X is a space, and that $\{X_s\}_{s \in S}$ is a collection of closed subspaces of X. Let $S(x) = \{s \in S \mid x \in X_s\}$. (We may think of $S(x)$ as the set of generators which fix x.) Define an equivalence relation $\sim$ on $W \times X$ by $(w, x) \sim (w', x')$ if and only if $x = x'$ and $w^{-1}w' \in W_{S(x)}$. Denote the quotient space $(W \times X)/ \sim$ by $\mathcal{U}(S, X)$. Thus, roughly speaking, $\mathcal{U}(S, X)$ is obtained by pasting together copies of X along the subspaces X_s in a fashion which depends upon the structure of W.

For instance, the Davis complex $\Sigma(W, S)$ is W-equivariantly homeomorphic to the space $\mathcal{U}(S, K)$, where K is the fundamental chamber for the action of W on Σ and K_s is the mirror corresponding to $s \in S$.

We can apply this same construction to the set D of Σ, relative to the closed subsets $D_{s'} = D \cap \Sigma^{s'}$, $s' \in S'$. Denote by $\mathcal{U}$ the space $\mathcal{U}(S', D)$ so defined. We have the following result:

Proposition 4.6   *The inclusion of D into Σ extends to a W-equivariant homeomorphism $f : \mathcal{U} \to \Sigma$. Moreover, f is a homotopy equivalence.*

In fact, $\mathcal{U}$ is a branched covering of Σ, as indicated by the following result.

Proposition 4.7   *Let $\Sigma^{(2)}$ denote the union of the codimension-2 subspaces of Σ. The map f restricted to $f^{-1}(\Sigma \setminus \Sigma^{(2)})$ is a covering projection, and the number of sheets containing any point is equal to the number of chambers in D.*

This fact should not be too surprising, because of the way in which $\mathcal{U}$ was constructed (*i.e.*, if there are p chambers in D, the map f provides a "p-fold covering" of Σ).

For a point x in the interior U of the fundamental chamber K, we can take this interior as a neighborhood U_x of x which is evenly covered by the map f (this choice can be made in general for $x \in U$). If $x \in \Sigma \setminus \Sigma^{(2)}$ but $x \notin \text{int}(K)$, x lies in some mirror K_s, $s \in S$. We may select an open subset, U_x, of K which contains x and which does not intersect any mirror K_t, $t \neq s$. It can then be shown that the set $U_x \cup sU_x$ is an open subset of Σ which is evenly covered by the map f. In either case, the preimage $f^{-1}(U_x)$ has the same number of sheets as D has chambers of Σ. Evenly covered neighborhoods of points in wK, $w \in W$, are constructed in the obvious fashion.

What remains is a bit of homological diagram chasing. For the remainder of this section, let U denote the interior of K, and denote by $U_1, ..., U_p$ the components of $f^{-1}(U)$.

Because Σ and $\mathcal{U}$ are contractible homology n-manifolds, we can define an orientation on each of the spaces K, Σ, D, and $\mathcal{U}$, by specifying an orientation of each n-simplex in the respective space as follows. First orient each n-simplex in K by selecting an orientation of K. Then each n-simplex $w\sigma$ in the translate wK inherits an orientation which is $(-1)^{l(w)}$ times the orientation on σ (where $l(w)$ is the length of w with respect to S, defined as in 1.3.2). This defines an orientation on Σ and on D; $\mathcal{U}$ can then be oriented by orienting each simplex $w\sigma$ in the translate wD by $(-1)^{l'(w)}$ times the orientation on σ in D, where $l'(w)$ is the length of w with respect to the generating set S'.

These orientations provide us with distinguished generators for each of the (infinite cyclic) groups $H^n(K, \partial K)$, $H^n_c(\Sigma)$, $H^n(\bar{U}_i, \partial \bar{U}_i)$, and $H^n_c(\mathcal{U})$, where $\bar{U}_i$ denotes the closure of U_i. (Note that each of these groups is indeed infinite cyclic because W is of type HM_n.) In particular, we denote by γ_K, γ_Σ, γ_i, and $\gamma_{\mathcal{U}}$ the class in each of these groups (respectively) dual to the orientation chosen in the previous paragraph.

By excision, the group $H^n(\Sigma, \Sigma \setminus U)$ can be identified with $H^n(K, \partial K) \cong \mathbb{Z}$, and similarly $H^n(\mathcal{U}, \mathcal{U} \setminus f^{-1}(U))$ can be identified with $\bigoplus H^n(\bar{U}_i, \partial \bar{U}_i) \cong \mathbb{Z}^p$. The existence of the following long exact sequence is guaranteed by basic homology theory:

$$\cdots \longrightarrow H^{n-1}_c(\Sigma \setminus U) \xrightarrow{\omega} H^n_c(\Sigma, \Sigma \setminus U) \xrightarrow{k^*} H^n_c(\Sigma) \xrightarrow{i^*} H^n_c(\Sigma \setminus U) \longrightarrow \cdots .$$
$$(4.4)$$

Since $H^n_c(K, \partial K) = H^n_c(\Sigma, \Sigma \setminus U)$, (4.4) shows that the class γ_K in $H^n(K, \partial K)$ gets taken to the class γ_Σ in $H^n_c(\Sigma)$ by the map k^*. Letting $U' = \cup_{i=1}^p U_i$, the exact sequence

$$\cdots \longrightarrow H^{n-1}_c(\mathcal{U} \setminus U') \xrightarrow{\omega} H^n_c(\mathcal{U}, \mathcal{U} \setminus U') \xrightarrow{k^*} H^n_c(\mathcal{U}) \xrightarrow{i^*} H^n_c(\mathcal{U} \setminus U') \longrightarrow \cdots$$
$$(4.5)$$

shows that the class γ_i of $H^n(\bar{U}_i, \partial \bar{U}_i)$ dual to the chosen orientation gets taken to the class $\gamma_{\mathcal{U}}$ in $H^n_c(\mathcal{U})$ by the natural map from $\bigoplus H^n(\bar{U}_i, \partial \bar{U}_i)$ to $H^n_c(\mathcal{U})$. Finally, the map

$$(f|U_i)^* : H^n(K, \partial K) \to H^n(\bar{U}_i, \partial \bar{U}_i)$$

induced by the restriction of f to U_i takes γ_K to γ_i. Applying the commutative diagram

$$\begin{array}{ccc} H^n(K, \partial K) & \longrightarrow & \bigoplus H^n(\bar{U}_i, \partial \bar{U}_i) \\ \downarrow & & \downarrow \\ H^n_c(\Sigma) & \longrightarrow & H^n_c(\mathcal{U}) \end{array}$$
$$(4.6)$$

to the element $\gamma_K \in H^n(K, \partial K)$, we obtain $f^*(\gamma_\Sigma) = p\gamma_{\mathcal{U}}$.

But because the map f is a proper W-equivariant homotopy equivalence, the degree of f must be 1. This forces $p = 1$, so that D indeed consists of a single chamber (because D had as many chambers as there were components in the preimage $f^{-1}(U))$. This completes the proof of strong rigidity in case W is of type HM_n.

4.3.5   Aut(W)

Suppose that W is strongly rigid, and moreover that for any given fundamental generating sets S and S' there is a unique word $w = w(S, S') \in W$ such that $wSw^{-1} = S'$. (That this condition is true in case W is of type HM_n or PM_n can be proven by an argument similar to the proof of Proposition 4.5.) Let (W, S) be a given system, and suppose $\alpha \in \text{Aut}(W)$, so that $(W, \alpha(S))$ is another system for W. Our hypotheses imply that $\alpha(S) = wSw^{-1}$ for some $w \in W$, so α can be written uniquely as a composition

$$\alpha = \beta \circ \gamma,$$

where $\beta \in \text{Inn}(W)$ and $\gamma \in \text{Diag}(W)$, proving

Theorem 4.6   *If W is of type HM_n or PM_n, then $\text{Aut}(W)$ is a semidirect product of $\text{Diag}(W)$ with $\text{Inn}(W)$.*

In particular, $|\text{Out}(W)| < \infty$. As we will see in Chapter 7, the basic semidirect product structure for $\text{Aut}(W)$ exhibited above will remain in more general settings, even when the group of outer automorphisms fails to be finite.

4.4   Two-dimensional groups

Instead of imposing a "manifold-like" structure upon $\Sigma(W, S)$, we can otherwise demand that its structure be simple by restricting our attention to certain two-dimensional Coxeter systems, as A. Kaul does in [Kaul (2000)]. As in 3.2.2, we say that a system (W, S) is of type K_n provided the corresponding diagram $\mathcal{V}$ is a complete graph on n vertices and every edge label is odd.

Theorem 4.7   *Suppose that (W, S) is of type K_n and that furthermore either all or all but one of the edges of the diagram $\mathcal{V}$ bear the same label. Then W is rigid.*

Remark 4.1   It follows at once from Theorem 3.13 (which we will prove independently) that groups of type K_n are strongly rigid. However, the special case covered by Theorem 4.7 can be obtained very easily from the work we have done already.

Let (W, S) be of type K_n. By Lemma 1.1, the nerve $N(W, S)$ is also a complete graph on n vertices. The first step in the proof of Theorem 4.7 is a refinement of Proposition 4.1:

Lemma 4.8 *Let (W, S) be of type K_n, and let (W, S') be another system for W, with diagram $\mathcal{V}'$. Then every maximal spherical simplex in $\mathcal{V}'$ is of rank 2. (From Proposition 4.1 it then follows that edges of $\mathcal{V}$ are conjugate to edges of $\mathcal{V}'$.)*

Proof. It is shown in [Charney and Davis (2000)] that if the sets $R(S)$ and $R(S')$ of reflections are equal, then for every positive integer k the sets of parabolic subgroups of rank k in each system are identical (see the proof of Proposition 4.4). Thus we are done if we can show $R(S) = R(S')$.

This is clear in case $n \leq 2$. Let $n \geq 3$ and suppose wsw^{-1} is a reflection in (W, S). Then

$$\{1, wsw^{-1}\} = w \left(\bigcap_{s \in T} W_T \right) w^{-1} = \bigcap_{s \in T} w W_T w^{-1}.$$

By Proposition 4.1, each group $w W_T w^{-1}$ is a parabolic subgroup of (W, S'), and therefore their intersection is a parabolic subgroup. (This is an elementary result if $w = 1$; otherwise, see [Charney and Davis (2000)].) That is, wsw^{-1} is a reflection in (W, S') as well.

The other inclusion $((R(S') \subseteq R(S))$ is only slightly more difficult (since we have no *a priori* knowledge of the rank of the parabolic subgroups of (W, S')). Given a reflection $ws'w^{-1} \in R(S')$, we may choose a maximal spherical simplex $\sigma' \subseteq S'$ containing s', so that $\{1, ws'w^{-1}\} \leq w W_{\sigma'} w^{-1}$. There is then a maximal spherical simplex $\sigma \subseteq S$ such that $v W_\sigma v^{-1} = W_{\sigma'}$, for some $v \in W$. It follows at once (from the fact that W_σ is a dihedral group of order $2l$ for some odd l) that

$$s' \in W_{\sigma'} = w^{-1} v W_\sigma v^{-1} w$$

is a reflection in (W, S). $\qquad \Box$

Let (W, S) be of type K_n, and let (W, S') be another system for W. It is now easy to show that every edge in $\mathcal{V}'$ has an odd label, and that no product $s't'$ has infinite order (for $s', t' \in S'$). Thus $\mathcal{V}'$ is a complete graph on n vertices, and the edge label multisets for both $\mathcal{V}$ and $\mathcal{V}'$ are identical.

If all but at most one of the edges carry the same label, the diagrams $\mathcal{V}$ and $\mathcal{V}'$ are obviously isomorphic.

4.5 Exercises

1. Show that if (W, S) is right-angled and (W, S') is another system for W then (W, S') is right-angled as well, and that $|S| = |S'|$.

2. Fill in the details in the proof of Proposition 4.1.

3. Prove Proposition 4.2.

4. Prove Lemma 4.2 and the first part of Lemma 4.3.

5. Fill in the gaps in the proof of Theorem 4.3.

6. Let A be a simplicial complex, and let σ be a k-simplex in A. Prove that the link $\mathrm{Lk}(\sigma, A)$ is an abstract simplicial complex. Moreover, verify that for any τ in A properly containing σ, the dimension of the simplex in $\mathrm{Lk}(\sigma, A)$ corresponding to τ has dimension $|\tau| - |\sigma| - 1$.

7. Prove that every Coxeter group of type HM_n is of type PM_n.

8. Let X be a pseudo-n-manifold. Prove that the link of any simplex in X is also a pseudo-manifold of the appropriate dimension, and that such a link is orientable if X is.

9. Fill in the details of the proof of Lemma 4.7. (Specifically, justify the claims made in the last sentence of the proof.)

10. (For those with a little knowledge of algebraic topology.) Let X be a space and let $\{X_s\}_{s \in S}$ be a collection of closed subspaces of X. Define $\mathcal{U}(S, X)$ as in 4.3.4. Prove the following two facts:

     a. $\mathcal{U}(S, X)$ is connected if and only if X is connected and $X_s \neq \emptyset$ for all $s \in S$.

     b. $\mathcal{U}(S, X)$ is simply connected if and only if X is simply connected, $X_s \neq \emptyset$ is connected for all $s \in S$, and $X_s \cap X_t \neq \emptyset$ if $m_{st} < \infty$.

Conclude that $\Sigma = \Sigma(W, S)$ is a connected and simply connected topological space.

Chapter 5

Even Coxeter groups

As was mentioned in Section 3.3, there are a number of techniques which can be applied to even Coxeter groups which do not apply to Coxeter groups in general. Geodesic words have a more regular structure and combinatorial objects such as centralizers and normalizers are much easier to describe. (This simpler structure is to a large extent a consequence of the fact that in an even system, distinct generators lie in distinct conjugacy classes.) Furthermore, the simple fact that 2 is an even number allows us to define a retraction from an even Coxeter group W onto any standard parabolic subgroup (as in Propositions 5.1 below). All of these facts facilitate study of the structure even Coxeter groups.

5.1 The structure of even Coxeter groups

We begin by proving some results which underscore the desirable structure of even Coxeter groups, beginning with the proposition below. Recall that a *retraction* $\rho : G \to H$ of a group G onto a subgroup H is a homomorphism such that $\rho|_H$ is the identity on H.

Proposition 5.1 *Let (W, S) be an even Coxeter system and let $T \subseteq S$. Then there is a retraction $\rho : W \to W_T$.*

Proof. Define ρ on the generating set S by letting $\rho(s) = s$ if $s \in T$, and $\rho(s) = 1$ otherwise. Because ρ so defined preserves each of the defining relations in the presentation corresponding to (W, S) (that is, $\rho\left((s_i s_j)^{m_{ij}}\right) = 1$), it is a homomorphism. By definition it is the identity on W_T. $\qquad\square$

It is easy to understand the action of ρ on a given word w (where w is a word in the generators of the even system): $\rho(w)$ is obtained by merely deleting from w all letters which do not lie in T.

Remark 5.1 We can in fact make a more general statement than that of Proposition 5.1. For a subgroup H of W, denote by $N(H)$ the normal closure of H in W (that is, $N(H)$ is the smallest normal subgroup of W containing H.)

Proposition 5.2 *Let (W, S) be any Coxeter system with diagram $\mathcal{V}$ and let $T \subseteq S$. Then $W/N(W_{S \setminus T})$ is isomorphic to the parabolic subgroup $W_{T_1} \leq W$, where $T_1 \subseteq T$ consists of generators s for which there is no path in $\mathcal{V}$ from s to an element of $S \setminus T$, every edge of which has an odd label.*

When (W, S) is even, this proposition clearly reduces to Proposition 5.1.

An immediate consequence of Proposition 5.1 (which goes hand-in-hand with the previous remark) is the following:

Corollary 5.1 *Let (W, S) be an even system, and let $s \neq t$ be generators in S. Then s and t are not conjugate to one another.*

Proof. Let ρ be the retraction onto the subgroup W_s. Then $\rho(s) = s$, whereas $\rho(wtw^{-1}) = 1$ for any conjugate of t. $\square$

Indeed, it is not difficult to see that two generators, s and t, are conjugate to one another if and only if there is a path in $\mathcal{V}$ from s to t for which every edge in the path has an odd label. The following result (from [Brady, *et al.* (2002)]) summarizes the structure of the conjugacy classes of W's reflections (even when (W, S) is not even). Its proof is left as an easy exercise.

Proposition 5.3 *Let (W, S) be an arbitrary Coxeter system, with diagram $\mathcal{V}$. Then the conjugacy classes of the reflections $R(S)$ are in one-to-one correspondence with the connected components of the diagram $\mathcal{V}_{\text{odd}}$ obtained from $\mathcal{V}$ by removing all edges with even labels.*

The following related result will prove useful later (in proving Theorem 3.8):

Proposition 5.4 *Suppose (W, S) and (W, S') are Coxeter systems such that $S \subseteq R(S')$. Then $R(S) = R(S')$.*

Proof. Exercise. $\square$

With Proposition 5.1 we may also prove the following useful lemma, which is key in many of the results which appear in this chapter.

Lemma 5.1 *Let (W, S) and (W, S') be even systems with corresponding diagrams $\mathcal{V}$ and $\mathcal{V}'$. Let σ_i ($i \in \{1, ..., n\}$) be simplices in $\mathcal{V}$, and let $\sigma = \bigcap_{i=1}^{n} \sigma_i$. Suppose that for every i, σ_i' is a simplex in $\mathcal{V}'$ such that for some $w_i \in W$, $W_{\sigma_i} = w_i W_{\sigma_i'} w_i^{-1}$. Let $\sigma' = \bigcap_{i=1}^{n} \sigma_i'$. Then there exists $w \in W$ such that $W_\sigma = w W_{\sigma'} w^{-1}$.*

Put simply, intersections of conjugate simplices are conjugate. Therefore once we are able to identify a class of simplices of $\mathcal{V}$ which are conjugate to simplices of $\mathcal{V}'$, we can enrich this class by closing it under intersection. It turns out (see [Bahls (2002)], for example) that this enriched class of simplices contains sufficient information to describe the global structure of the group W when (W, S) is even.

Proof. We prove the lemma in case $n = 2$; an obvious induction then completes the proof.

Let σ_i, σ_i', σ, σ', and w_i be as above, and suppose that u is a word in W_σ. Then

$$u = w_i u_i w_i^{-1}, i = 1, 2,$$

for some $u_i \in W_{\sigma_i'}$, so that

$$u_2 = w' u_1 w'^{-1}$$

for $w' = w_2^{-1} w_1$. Let $\rho : W \to W_{S' \setminus (\sigma_1' \setminus \sigma')}$ be the retraction defined in the proof of Proposition 5.1. Then $\rho(u_2) = u_2$, but

$$\rho(u_2) = \rho(w')\rho(u_1)\rho(w')^{-1},$$

so u_2 and $\rho(u_1)$ are conjugate. Clearly $\rho(u_1)$ is in $W_{\sigma'}$. It follows that $u \in w W_{\sigma'} w^{-1}$, where $w = w_2 \rho(w')$, so $W_\sigma \subseteq w W_{\sigma'} w^{-1}$. Analogously, $W_{\sigma'} \subseteq \bar{w} W_\sigma \bar{w}^{-1}$ for some $\bar{w} \in W$.

If we now apply the retraction $\rho' : W \to W_\sigma$ to the chain

$$W_\sigma \subseteq w W_{\sigma'} w^{-1} \subseteq w \bar{w} W_\sigma \bar{w}^{-1} w^{-1}, \tag{5.1}$$

we note that the restriction of ρ' to $w \bar{w} W_\sigma \bar{w}^{-1} w^{-1}$ is an injection, giving an isomorphism to W_σ. Therefore the first term and the last term in (5.1) are equal to one another, and thus so is the middle term: $W_\sigma = w W_{\sigma'} w^{-1}$, as desired. $\square$

This is all well and good, but to what simplices can we apply this result? This is the focus of the next section.

5.2   Conjugate simplices in even systems

Let us suppose that we are given two even systems, (W, S) and (W, S'), with corresponding diagrams $\mathcal{V}$ and $\mathcal{V}'$. In order to apply Lemma 5.1, we must have in hand a collection of simplices of V which are conjugate to simplices of $\mathcal{V}'$. Such a collection is provided by the following result.

Proposition 5.5   *Let (W, S) and (W, S') be even systems, with diagrams $\mathcal{V}$ and $\mathcal{V}'$, respectively. If σ is a maximal spherical simplex in $\mathcal{V}$ (see 1.3.3), then there exists a maximal spherical simplex σ' in $\mathcal{V}'$ such that W_σ and $W_{\sigma'}$ are conjugate.*

Proof.   The proposition is any easy consequence of Lemma 1.2 and a "ping-pong" argument similar to that given for Lemma 5.1. The details are left as an exercise.   □

By applying Proposition 5.5 and Lemma 5.1 together, we obtain the following fact:

Corollary 5.2   *Let (W, S) and (W, S') be even systems, with diagrams $\mathcal{V}$ and $\mathcal{V}'$, respectively. If $\sigma \subseteq V$ is an intersection of maximal spherical simplices in $\mathcal{V}$, then there exists a simplex $\sigma' \subseteq \mathcal{V}'$ such that W_σ and $W_{\sigma'}$ are conjugate.*

Given systems (W, S) and (W, S'), we will denote by $\mathcal{M}(S)$ the set of maximal spherical simplices in $\mathcal{V}$, and by $\mathcal{K}(S, S')$ the set of simplices σ of V such that for some $\sigma' \subseteq \mathcal{V}'$, W_σ and $W_{\sigma'}$ are conjugate. (We use the letter 'K' rather than 'C', as the latter is reserved to denote centralizers, in other chapters.) In particular, $\mathcal{M}(S) \subseteq \mathcal{K}(S, S')$.

Remark 5.2   When our choices of S and S' are clear, we abbreviate: $\mathcal{M} = \mathcal{M}(S)$ and $\mathcal{K} = \mathcal{K}(S, S')$.

When (W, S) is even, the form of W_σ for $\sigma \in \mathcal{M}(S)$ is very straightforward. As every edge in $\mathcal{V}$ is even, any spherical subgroup of W can be expressed as a direct product of dihedral groups D_{2k} $(k \geq 2)$ and the 2-element group $\mathbb{Z}_2$ (Lemma 4.2). By Lemma 4.3, this expression can be "refined" only by writing D_{2k} as $D_k \times \mathbb{Z}_2$ when k is odd. This finest decomposition is essentially unique, according to the following general theorem, known as the Krull-Schmidt Theorem (see [Rotman (1995)]):

Theorem 5.1 *Suppose that a group G satisfies both the ascending and descending chain conditions for normal subgroups (in particular, finite groups clearly satisfy both of these properties). Given any two decompositions of G,*

$$G \cong H_1 \times H_2 \times \cdots \times H_m,$$

and

$$G \cong K_1 \times K_2 \times \cdots \times K_n,$$

into directly indecomposable groups H_i and K_j, $m = n$ and there is a permutation ϕ of $\{1, ..., m\}$ such that $H_i = K_{\phi(i)}$ for every $i = 1, ..., m$.

Now assume that $\sigma \in \mathcal{M}(S)$, and that (W, S') is another even system for W, and that $\sigma' \in \mathcal{M}(S')$ satisfies $W_\sigma = wW_{\sigma'}w^{-1}$, so that $W_\sigma \cong W_{\sigma'}$. Both W_σ and $W_{\sigma'}$ can be decomposed into direct products of indecomposable dihedral groups and $\mathbb{Z}_2$. Applying the Krull-Schmidt Theorem guarantees that both of these finest decompositions are identical. Since both decompositions come from even presentations, each "odd" dihedral factor in either decomposition is paired with one of the copies of $\mathbb{Z}_2$ without fail, and the "visual" decompositions of W_σ and $W_{\sigma'}$ are identical. That is, σ and σ' are isomorphic simplices.

Proposition 5.6 *If σ and σ' are even simplices and $W_\sigma \cong W_{\sigma'}$, then σ and σ' are isomorphic as edge-labeled graphs.*

We are now in a position to establish a pair of "weak" rigidity results for all even Coxeter groups. Here is the first:

Theorem 5.2 *If $\mathcal{V}$ and $\mathcal{V}'$ are even diagrams corresponding to the same Coxeter group W, then $\mathcal{V}$ and $\mathcal{V}'$ are isomorphic as unlabeled graphs.*

Proof. Let $\sigma \in \mathcal{M}(S)$ and let $\sigma' \in \mathcal{M}(S')$ be the corresponding simplex in $\mathcal{V}'$. Consider the commutator subgroups $H = [W_\sigma, W_\sigma]$ and $H' = [W_{\sigma'}, W_{\sigma'}]$. Because $W_\sigma = wW_{\sigma'}w^{-1}$, $N(H) = N(H')$. A diagram for the Coxeter group $W/N(H) = W/N(H')$ is obtained from $\mathcal{V}$ by changing the label on each edge of σ to 2. Similarly, a diagram for the same group is obtained from $\mathcal{V}'$ by changing the label on each edge of σ' to 2.

At this point we could induct on the "complexity" of $\mathcal{V}$ and $\mathcal{V}'$ (i.e., on the total number of edges with labels exceeding 2) to obtain the desired result, as is done in the proof of the next theorem. The base case, in which W is right angled, is given by Theorem 4.1. Alternatively, we could apply the procedure above to each element of $\mathcal{M}(S)$ in turn, yielding the same result.   □

Not only are the graphs underlying $\mathcal{V}$ and $\mathcal{V}'$ identical; the multisets of edge labels are the same:

Theorem 5.3   *Let $\mathcal{V}$ and $\mathcal{V}'$ be even diagrams for the same Coxeter group W. Then for any integer n, the number of edges of $\mathcal{V}$ with label n is equal to the number of edges of $\mathcal{V}'$ with label n.*

Proof.   Were the theorem false, we could choose a group W, diagrams $\mathcal{V}$ and $\mathcal{V}'$, and an even integer k such that the number of edges labeled k in $\mathcal{V}$ differs from the number of such edges in $\mathcal{V}'$, and the number of edges labeled k in $\mathcal{V}$ is minimal among all such choices of groups, diagrams, and integers.

Suppose now that the edge e in $\mathcal{V}$ has label k, and that e lies in $\sigma \in \mathcal{M}(S)$. Let $\sigma' \in \mathcal{M}(S')$ be such that W_σ and $W_{\sigma'}$ are conjugate. Thus σ and σ' contain the same (positive) number of edges labeled k. Let now $H = [W_\sigma, W_\sigma]$ and $H' = [W_{\sigma'}, W_{\sigma'}]$, so that $N(H) = N(H')$. As in the proof of Theorem 5.2, diagrams for $W/N(H) \cong W/N(H')$ are obtained from $\mathcal{V}$ and $\mathcal{V}'$ by relabeling all edges of σ and σ' with the number 2. Clearly then $W/N(H)$, with the new diagrams just constructed, and the number of edges labeled k in the first of these diagrams, gives a contradiction to the minimality of our original choice.   □

These results hint at the following theorem (paraphrased in Chapter 3 as Theorem 3.9), whose proof can be found in [Bahls (2002)]:

Theorem 5.4   *Let $\mathcal{V}$ and $\mathcal{V}'$ be even diagrams for the same Coxeter group, W. There is an edge-labeled graph isomorphism $\phi : \mathcal{V} \to \mathcal{V}'$ such that*

1. *If $\sigma \in \mathcal{K}(S, S')$ and $\sigma' \in \mathcal{K}(S', S)$ such that W_σ and $W_{\sigma'}$ are conjugate, then ϕ takes the collection of vertices of σ to the collection of vertices of σ'.*
2. *If $[st] \subseteq \mathcal{V}$ and $[s't'] \subseteq \mathcal{V}'$ are edges such that the groups $\langle (st)^2 \rangle$ and $\langle (s't')^2 \rangle$ are conjugate, then $\{\phi(s), \phi(t)\} = \{s', t'\}$.*

That is, there is at most one even diagram (up to edge-labeled graph isomorphism) for any given Coxeter group. This fact should not be overly surprising, for Theorems 5.2 and 5.3 imply that the only way in which two diagrams, $\mathcal{V}$ and $\mathcal{V}'$ could differ is in the way in which the maximal spherical simplices intersect (their "relative orientations").

Theorem 5.4 was originally proven using a construction called the *pattern* of a Coxeter system. However, the theorem also follows naturally from the method of centralizer chasing, introduced in Section 6.2.

We will return to address some consequences of Theorem 5.4 after discussing a few other rigidity results which involve reflection independence.

5.3   Reflection rigidity in even groups

In the next section we will develop a method of determining whether or not an even Coxeter system is reflection preserving or reflection independent. In this section, we assume that we are given two systems, (W, S) and (W, S'), one of which (the first, say) is even, satisfying $R(S) = R(S')$, and show that the corresponding diagrams must be isomorphic. That is, we have the following result:

Theorem 5.5   *Let (W, S) be an even system. Then (W, S) is reflection rigid.*

We follow the proof of [Brady, *et al.* (2002)], which in turn builds upon [Dyer (1990)].

Theorem 5.5 follows almost at once from the following fact:

Theorem 5.6   *Let (W, S) and (W, S') be Coxeter systems such that $R(S) = R(S')$. Then $|S| = |S'|$.*

Let us assume Theorem 5.6 and prove Theorem 5.5. We have (W, S) and (W, S'), where the first system is even, and by hypothesis $R(S) = R(S')$ (so that $|S| = |S'|$). By Proposition 5.3, using the notation of that result, the connected components of $\bar{\mathcal{V}}'$ are in one-to-one correspondence with the conjugacy classes of S. Since $|S| = |S'|$, no two vertices of $\mathcal{V}$ lie in the same conjugacy class, so every edge of $\mathcal{V}'$ must be even.

Consider vertices $s \neq t$ in $\mathcal{V}$, and let C_s and C_t be the conjugacy classes of these generators in W. There are unique vertices, s' and t', in $\mathcal{V}'$, which lie in C_s and C_t, respectively. Suppose that for some $c_s \in C_s$ and $c_t \in C_t$, $\langle c_s, c_t \rangle$ is a finite dihedral group, G. By Lemma 1.2, G can be conjugated inside $W_{\{s,t\}}$ on the one hand, and inside $W_{\{s',t'\}}$ on the other. The groups $W_{\{s,t\}}$ and $W_{\{s',t'\}}$ thus contain the same dihedral subgroups, and must therefore be isomorphic. Thus there is an edge $[st]$ in $\mathcal{V}$ if and only if there is an edge $[s't']$ in $\mathcal{V}'$, and these edges (if present) must bear the same labels. This is enough to show that $\mathcal{V}$ and $\mathcal{V}'$ are isomorphic, as edge-labeled graphs.

So we must now prove Theorem 5.6. We indicate the important steps, as they are taken in [Dyer (1990)]. Many of the intermediate results indicate an interesting generalization of Coxeter groups which we have not yet considered (namely, that of a reflection system, defined below).

The following theorem is crucial:

Theorem 5.7 *Let (W, S) be a Coxeter system, and let $R' \subseteq R(S)$. Then the subgroup G of W generated by R' is a Coxeter group with a fundamental generating set S' such that $|S'| \leq |R'|$.*

In proving this theorem, one may obtain a canonical description of the generating set S'; in case $G = W$, S' turns out to be S itself. Applying Theorem 5.7 to the case in which $R' = S'$, we obtain $|S| \leq |S'|$. Symmetrically, $|S'| \leq |S|$, so the sets have equal size. Thus Theorem 5.6 will be proven.

The statement of Theorem 5.7 makes sense in a more general setting. Suppose that G is a group with generating set $X \subseteq G \setminus \{1\}$. Define $R(X)$ to be the set of conjugates of elements of X:

$$R(X) = \bigcup_{g \in G} gXg^{-1}. \tag{5.2}$$

Let $o(r)$ denote the order of the group element r, for each $r \in R(X)$. Let $\mathbb{Z}(r) = \mathbb{Z}_{o(r)}$ for every r (where $\mathbb{Z}_n$ is the cyclic group of order n, and $\mathbb{Z}_\infty = \mathbb{Z}$). We define the $\mathbb{Z}G$-module $M(G, X)$ to be the set of formal sums

$$\left\{ \sum_{r \in R(X)} a_r r \mid a_r \in \mathbb{Z}(r) \text{ for all } r \in R(X), a_r = 0 \text{ for almost all } r \right\}. \tag{5.3}$$

G then acts on $M(G, X)$ by

$$g \cdot \sum a_r r = \sum a_r (grg^{-1}). \tag{5.4}$$

We say that (G, X) is a *reflection system* if there is a map $N : G \to M(G, X)$ satisfying

1. $N(gh) = N(g) + g \cdot N(h)$ for all $g, h \in G$, and
2. $N(x) = 1x$ for all $x \in X$.

Such a map is called a *reflection cocycle*, if it exists.

Remark 5.3 Any Coxeter system (W, S) is a reflection system, with N defined by $N(w) = \sum_{R_w} r$, where $R_w = \{r \in R \mid l(rw) < l(w)\}$. (Therefore anything we prove for reflection systems will be true about Coxeter systems.) Roughly speaking, the set R_w is the set of reflections which cannot form a prefix of any reduced expression representing the group element w. The set $\{s \in S \mid l(sw) < l(w)\}$ is closely related to the set W^T defined in (1.8). This set is also related to the *Bruhat order*, a partial order (defined on the elements of W) which is compatible with the length function.

It is possible to show that if (G, X) is a reflection system and X consists entirely of involutions (*i.e.*, elements of order 2), then W is a Coxeter group, and (W, S) is a Coxeter system. (See Theorem 1.3.)

Much as the length of an element $w \in W$ in a Coxeter system is defined in terms of the generating set S, we define the length $l(g)$ of an element $g \in G$ to be the smallest value of $n \in \mathbb{Z}$ such that $g =_G x_1 \cdots x_n$, where $x_i \in X \cup X^{-1}$ for all $i = 1, ..., n$. With this definition in hand, we can state the following lemmas. The first should remind the reader of the Exchange Condition (see 1.3.4), while the second essentially says that "reflections must look like reflections".

Lemma 5.2 *Let (G, X) be a reflection system. Suppose $g = x_1^{n_1} \cdots x_m^{n_m}$, $r \in R(X)$, and $\epsilon \in \{\pm 1\}$ are given so that $l(r^\epsilon g) \leq l(g)$. Then for some index i, $1 \leq i \leq m$,*

$$r = (x_1^{n_1} \cdots x_{i-1}^{n_{i-1}}) x_i (x_{i-1}^{-n_{i-1}} \cdots x_1^{-n_1}),$$

and

$$r^\epsilon g = x_1^{n_1} \cdots x_{i-1}^{n_{i-1}} x_i^{n_i + \epsilon} x_{i+1}^{n_{i+1}} \cdots x_m^{n_m}.$$

Lemma 5.3 *Let (G, X) be a reflection system, and let $r \in R(X)$. Then $l(r)$ is odd, and if $r = x_1^{\epsilon_1} \cdots x_{2m+1}^{\epsilon_{2m+1}}$ for some $x_i \in X$ and $\epsilon_i \in \{\pm 1\}$, then*

$$r = (x_1^{\epsilon_1} \cdots x_m^{\epsilon_m}) x_{m+1}^{\epsilon_{m+1}} (x_m^{-\epsilon_m} \cdots x_1^{-\epsilon_1}).$$

There are further similarities with Coxeter groups. In fact, given a reflection system (G, X), one can define a graph which completely describes the structure of the system, much as a Coxeter graph describes the structure of the corresponding Coxeter system. (The major difference is that now generators may have orders other than 2, so that vertices of the representative graph must also be labeled.)

Now let us assume that we have in hand a reflection system, (G, X), and let H be any subgroup of G. Let $R = R(X)$, and define R_H to be the set of reflections which lie in H: $R_H = R(X) \cap H$. If H satisfies $H = \langle R_H \rangle$, we call H a *reflection subgroup* of G. Given any element $a = \sum_{r \in R} a_r r$ of the $\mathbb{Z}G$-module $M(G, X)$, we may "restrict" a to the set R_H, defining $a_H = \sum_{r \in R_H} a_r r$.

Let H be a reflection subgroup of G, and define

$$\chi(H) = \{r \in R \mid N(r)_H = 1r\}.$$

This operator will be used to construct the canonical generating set for H which appears in Theorem 5.7.

Remark 5.4 We may gain intuition by recalling the case of a Coxeter system (W, S). Here, roughly speaking, a reflection r lies in $\chi(H)$ if and only if r is the only reflection with which a reduced expression for r may begin.

It is not hard to see that $\chi(H) \subseteq H$ and $X \cap H = X \cap \chi(H)$ both hold for any reflection subgroup H.

Proposition 5.7 *Let (G, X) be a reflection system, and let H be a reflection subgroup of G. Then the following are true.*

1. *Let $r \in R_H$. Then $r = (r_1^{\epsilon_1} \cdots r_m^{\epsilon_m}) r_0 (r_m^{-\epsilon_m} \cdots r_1^{-\epsilon_1})$ for some $m \geq 0$, $r_i \in \chi(H)$, and $\epsilon_i \in \{\pm 1\}$. (In particular, $\chi(H)$ generates H.)*
2. *Define $N_H : H \to M(G, X)$ by $N_H(g) = N(g)_H$. Then for all $g \in G$ and $h \in H$, $N_H(hg) = N_H(h) + h \cdot N_H(g)$. Furthermore, $(H, X \cap H)$ is a reflection system, with reflection cocycle given by N_H.*

We can now describe the process by means of which a canonical generating set S' is obtained from R'. Let (G, X) be a reflection system with set of reflections R. Suppose that $R' \subseteq R$ is given, and that G' is the subgroup of G generated by R. Define R_i $(i \geq 0)$ as follows:

1. $R_0 = R'$,
2. $R_{i+1} = R_i$ if $\chi(\langle r, r' \rangle) = \{r, r'\}$ for all $r \neq r'$ in R_i. Otherwise, let $R_{i+1} = (R_i \setminus \{r, r'\}) \cup \chi(\langle r, r' \rangle)$ for some pair $\{r, r'\}$ satisfying $\chi(\langle r, r' \rangle) \neq \{r, r'\}$.

We then have:

Proposition 5.8 *For some i, $R_{i+1} = R_i$ holds, and for this i, $\chi(G') = R_i$. Moreover, $|R_i| \leq |R'|$.*

This proposition is proven by showing that if $r \neq r'$ lie in R with $G' = \langle r, r' \rangle$, then $|\chi(G')| = 2$, and if $\chi(G') = \{r_1, r_2\} \neq \{r, r'\}$, then $l(r_1) + l(r_2) < l(r) + l(r')$. The second fact enables us to induct upon the total length of the reflections in R_i, and the first guarantees that at each step, $|R_i|$ does not increase (proving that $|R_i| \leq |R'|$ for every i). Essentially, R_i is obtained from R' by exchanging one pair of reflections for another pair, perhaps many times.

Furthermore, in case $G' = W$, it can be shown that the two-step process given above yields the fundamental generating set originally given for the Coxeter group W. Theorem 5.7 follows, finally completing the proof of Theorem 5.5.

While we're at it, let us prove the remaining parts of Theorem 3.5, the first of which is almost immediate.

Assume that W is finite. It is a simple matter to reduce to the case in which both systems (W, S) and (W, S') are irreducible (see Exercise 6). By Theorem 5.6, $|S| = |S'|$, and we can appeal to Chapter 2 of [Radcliffe (2000)], which shows that no isomorphisms exist between two irreducible finite Coxeter groups of different type.

Now to the case in which $\mathcal{V}$ is a tree. The following lemma follows easily from the work we've now done:

Lemma 5.4 *Let (W, S) be a system with diagram $\mathcal{V}$ such that $\mathcal{V}$ is a finite tree, every edge of which has an odd label. Then (W, S) is rigid up to diagram twisting.*

Proof. Let (W, S') be another system for W. By Proposition 5.3, there is exactly one conjugacy class of generators (and therefore reflections) in (W, S); from this it follows that $\mathcal{V}'$ must be connected. It is easy to show (please do so!) that the reflections $R(S)$ are the only involutions of W, so that $R(S) = R(S')$. $|S| = |S'|$ follows from Theorem 5.6, so $\mathcal{V}'$ has the same number of vertices as $\mathcal{V}$. Moreover, the edge label multisets are identical, by Proposition 4.1. Therefore $\mathcal{V}'$ is a finite connected graph with the same number of edges as $\mathcal{V}$, and the same multiset of edge labels. Simple combinatorics forces $\mathcal{V}'$ to be a tree.

Now it is clear that, given any $s \in S$, we can twist any component of the full subgraph of $\mathcal{V}$ on the set $S \setminus \{s\}$ around s to obtain a new tree with the same multiset of edge labels. In particular, we can twist repeatedly until we obtain a "starlike" graph, in which only one vertex has valence exceeding 1. The same can be done with the diagram $\mathcal{V}'$, so that $\mathcal{V}$ and $\mathcal{V}'$ must be twist-equivalent. $\square$

For any diagram $\mathcal{V}$, we define $\mathcal{V}_{\text{odd}}$ to be the graph resulting from $\mathcal{V}$ by removing all edges with even labels (as in Proposition 5.3). As in [Brady, *et al.* (2002)], we define $\mathcal{V}_{\text{even}}$ to be the graph resulting from $\mathcal{V}$ by *contracting* every edge with an *odd* label.

Lemma 5.5   *Let $\mathcal{V}$ and $\mathcal{V}'$ be finite trees. Suppose that $\mathcal{V}_{\text{even}} \cong \mathcal{V}'_{\text{even}}$ by an isomorphism ϕ such that for every $s \in \mathcal{V}_{\text{even}}$, the component of s in $\mathcal{V}_{\text{odd}}$ has the same multiset of edge labels as the connected component of $\phi(s)$ in $\mathcal{V}_{\text{odd}}$. Then $\mathcal{V}'$ can be obtained from $\mathcal{V}$ by applying a sequence of diagram twists.*

Proof.   Exercise. (*Hint*: show that both are twist-equivalent to a diagram $\mathcal{V}''$ in which $\mathcal{V}''_{\text{even}}$ is a *subdiagram* of $\mathcal{V}''$.)   □

Finally, suppose that $\mathcal{V}$ is a finite tree with arbitrary edge labels. Let (W, S') be another system corresponding to W such that $R(S) = R(S')$. Applying Proposition 4.1, Theorem 5.3, and Theorem 5.6, $\mathcal{V}'$ must have the same number of vertices and edges as $\mathcal{V}$, and the edge label multisets must be the same. Another application of Proposition 4.1 shows that the edge label multiset of any component of $\mathcal{V}_{\text{odd}}$ is identical to the edge label multiset of some component $\mathcal{V}'_{\text{odd}}$ (how?). Moreover, applying the same proposition (yet again!) shows that if two components U_1 and U_2 in $\mathcal{V}_{\text{odd}}$ are connected by an even edge labeled m, then the components in $\mathcal{V}'_{\text{odd}}$ corresponding to U_1 and U_2 as in the previous sentence must also be connected by an edge labeled m.

In particular, $\mathcal{V}'$ must also be connected, and therefore a tree (as in the proof of Lemma 5.4). The arguments of the previous paragraph also show that $\mathcal{V}_{\text{even}} \cong \mathcal{V}'_{\text{even}}$, so that we can apply Lemma 5.5 to conclude the proof.

5.4   Reflection preservation and reflection independence

Now that we have in hand some consequences of reflection independence, let us discover when this phenomenon occurs.

The goal of this section is to sketch a proof of Theorem 3.8, which enables us to determine (merely by examining the corresponding diagram) whether or not a given even system is reflection preserving. Moreover, coupled with Theorem 5.5, we will be able to tell when a given even system is reflection independent. (If (W, S) is even and not rigid, then it cannot be reflection independent, because of Theorem 5.5.)

We first take care of the first two parts of Theorem 3.8. Much of the notation used throughout the remainder of this chapter was defined in Chapter 3. Also, for a generator $s \in S$, we denote by τ_s the following collection of vertices:

$$\bigcap \{\sigma \in \mathcal{M}(S) \mid s \in \sigma\}.$$

Proposition 5.9   *Let (W, S) be an even Coxeter system with diagram $\mathcal{V}$. Suppose that one of the following conditions is satisfied (see Figure 3.3):*

1. *there exist vertices $s \neq t$ in $\mathcal{V}$ such that $\mathrm{St}(s) \subseteq \mathrm{St}_2(t)$, or*
2. *there exist distinct vertices s_1, s_2, and s_3 and an edge $[s_2 s_3]$ with label $n > 2$ in $\mathcal{V}$ such that both s_2 and s_3 lie in τ_{s_1}.*

Then (W, S) is neither reflection preserving nor reflection independent.

Proof.   Suppose the first condition is satisfied. Define $\alpha : W \to W$ by $\alpha(s) = st$. It is easy to check that all of the relations of the presentation corresponding to (W, S) are preserved, so that α is a homomorphism. Moreover, α is an involution, so that it is an automorphism of W. Since st has even length, it is not a reflection with respect to (W, S), so (W, S) and $(W, \alpha(S))$ are systems with different reflections.

Suppose now that the second condition is satisfied. Define $\alpha : W \to W$ by $\alpha(s_1) = s_1(s_2 s_3)^{n/2}$. As before, is easy to check that α is a homomorphism. (Note that by Lemma 1.1, s_1 must commute with both s_2 and s_3, and if $s_1 t$ has order greater than 2, then both s_2 and s_3 commute with t.) Once more, α is an involution, so $\alpha \in \mathrm{Aut}(W)$. Moreover, applying Lemma 5.3, we see that $\alpha(s_1)$ is not a reflection, so (W, S) and $(W, \alpha(S))$ have different reflections. $\square$

Now we assume that neither condition in Proposition 5.9 holds. By Proposition 5.4, we need only show that $S \subseteq R(S')$ in order to prove Theorem 3.8. If this were not so, we could choose a generator $s \in S \setminus R(S')$ satisfying the following minimality requirement:

$$t \in S \setminus R(S') \Rightarrow |\tau_t| \geq |\tau_s|. \tag{5.5}$$

Now we consider two cases, depending on whether W_{τ_s} is abelian or not. The following lemma will be useful:

Lemma 5.6   *Let $\sigma_i \in \mathcal{M}(S)$ and $\sigma_i' \in \mathcal{M}(S')$, and let $W_{\sigma_i} = w_i W_{\sigma_i'} w_i^{-1}$ for all $i = 1, ..., k$. Let $\sigma = \cap \sigma_i$ and $\sigma' = \cap \sigma_i'$, and suppose $W_\sigma = w W_{\sigma'} w^{-1}$. Assume that W_σ is right-angled, and let $s \in \sigma$. Then if $s = v_1 u_1 v_1^{-1} =$*

$v_2 u_2 v_2^{-1}$ *for geodesics u_i in the letters of σ' and $v_i \in W$, then $u_1 =_W u_2$.*
Furthermore, if s is central in W_{σ_i}, then every letter of u_i is central in $W_{\sigma'_i}$.

Proof.   This can be proven using quotient maps which identify to 1 the appropriate generators of S'. The details are left as an exercise.   □

Case 1.   Suppose W_{τ_s} is abelian. Since $\tau_s \in \mathcal{K}(S, S')$, we may choose $\tau' \subseteq \mathcal{V}'$ and $w \in W$ such that $W_{\tau_s} = wW_{\tau'}w^{-1}$. In particular, $s = ws'_1 \cdots s'_k w^{-1}$ for some $s'_i \in \tau'$, $k \geq 2$.

Lemma 5.7   *For some i, $1 \leq i \leq k$, $\tau' = \tau_{s'_i}$.*

Proof.   If not, then for each i we can choose a proper subset $\tau_i \subsetneq \tau_s$ such that W_{τ_i} and $W_{\tau_{s'_i}}$ are conjugate. Moreover, $s \notin \tau_i$, for all i. Therefore, by the minimality condition (5.5), there exist $s_i \in \tau_i \subsetneq \tau_s$ and $w_i \in W$ such that $s_i = w_i s'_i w_i^{-1}$, for all i. But $s_i \in wW_{\tau'}w^{-1}$ for every i, so that $s_i = wu_i w^{-1}$ for some word u_i in the letters of τ'. Applying Lemma 5.6, we obtain $s_i = ws'_i w^{-1}$, so

$$s = ws'_1 \cdots s'_k w^{-1} = s_1 s_2 \cdots s_k,$$

which is ridiculous.   □

Because τ' is abelian, we may assume that $\tau' = \tau_{s'_1}$. Given a group G and $g \in G$, let $N(g)$ denote the normal closure of g in G. We now construct two diagrams for the Coxeter group $W/N(s)$, whose structures will lead to a contradiction to the assumption that $s \notin R(S')$.

Lemma 5.8   *With the above notation, new diagrams for $W/N(s) = W/N(s'_1 \cdots s'_k)$ can be obtained in either of the following ways:*

1. from $\mathcal{V}$ by removing s and all incident edges, and
2. from $\mathcal{V}'$ by removing s'_1 and all incident edges, and relabeling each edge $[ts'_i]$ to 2, where $t \in \mathrm{Lk}(s'_1)$ and $2 \leq i \leq k$.

Proof.   The first statement is obvious. To prove the second, we show that in replacing s'_1 by the word $s'_2 \cdots s'_k$ and then removing s'_1 itself (and perhaps adding some new relations), we obtain a presentation which corresponds to the diagram described in (2) above.

The relations $(s'_2 \cdots s'_k)^2 = 1$ and $((s'_2 \cdots s'_k)s'_1)^2 = 1$ are clearly consequences of the relations not containing s'_1.

Suppose $(s'_1 t)^n = 1$ $(n > 2)$ is a relation in the original presentation. Since $\tau' = \tau_{s'_1}$, every s'_i is adjacent to t, and Lemma 1.1 implies that $[s'_i t]$

has label 2. Therefore $((s'_2 \cdots s'_k)t)^n = 1$ follows from the original relations not involving s'_1.

Suppose $(s'_1 t)^2 = 1$ for some $t \in S' \setminus \tau'$. Again, each s'_i is adjacent to t, and (using Lemma 1.1 again) at most one of the edges $[s'_i t]$ (say, $[s'_2 t]$) has label m greater than 2. Then the relation $((s'_2 \cdots s'_k)t)^2 = 1$ is a consequence of the original relations not involving s'_1 and the new relation $(s'_2 t)^2 = 1$, which displaces the old relation $(s'_2 t)^m = 1$. (Adding this last relation corresponds to changing to label on $[s'_2 t]$ from $m > 2$ to 2.) $\square$

The reader familiar with the notion of a *Tietze transformation* will notice that the above proof merely involves the application of various such transformations, precipitated by adding the new relation $s'_1 = s'_2 \cdots s'_k$.

For any element or subset A in W, let $\bar{A}$ denote the image of A in $W/N(s)$. By the first part of Lemma 5.8, each $\bar{s}'_i$ $(i \geq 2)$ is central in every $W_{\bar{\sigma}'}$, where $\sigma' \in \mathcal{M}(S')$ is conjugate to $\sigma \in \mathcal{M}(S)$ which contains s. This implies (by Lemma 5.6) that each $\bar{s}_i$ is central in $W_{\bar{\sigma}}$. By the second part of Lemma 5.8, each s_i appearing in the above product must be central in W_σ. This holds for *every* $\sigma \in \mathcal{M}(S)$ containing s. But since $\mathrm{St}(s) \not\subseteq \mathrm{St}_2(s_i)$, this cannot be.

Case 2. Now assume that W_{τ_s} is not abelian. There is a unique edge, $[st]$, in τ_s which has label n, $n > 2$ (why?).

We may argue much as was done in the proof of Lemma 5.8. The main course of the argument is outlined below, while the details are left to the reader.

As before, let $\tau' \in \mathcal{K}(S', S)$ satisfy $W_{\tau_s} = wW_{\tau'}w^{-1}$. By Proposition 5.6, τ' has a unique edge, say $[s't']$, with label n, $n > 2$. Denote by S_1 (respectively, S'_1) the set of generators $\tau \setminus \{s, t\}$ (respectively, $\tau' \setminus \{s', t'\}$). Then both W_{S_1} and $W_{S'_1}$ are abelian, and it follows that for $v \in S_1$, $\tau_v \subseteq \tau_s$ is abelian. Therefore if $v \in S_1$, there exists a unique $v' \in S'_1$ such that v and v' are conjugate.

Keeping in mind that neither s nor t is central in W_{τ_s}, the following fact is now not difficult to prove:

Lemma 5.9 *For some reflections r_1 and r_2 in $W_{\{s',t'\}}$, and some w_1 and w_2 in $W_{S'_1}$, $s = wr_1w_1w^{-1}$ and $t = wr_2w_2w^{-1}$. Furthermore, up to relabeling of $[st]$, we may suppose that the central letter of r_1 (respectively, r_2) is s' (respectively, t').*

As in Lemma 5.8, we obtain a diagram for $W/N(s) = W/N(s'w_1)$ by removing from $\mathcal{V}$ the vertex s and all incident edges. Furthermore, we obtain a second diagram by modifying the presentation $\langle S' \mid \mathcal{R}' \rangle$ corresponding to (W, S'). Namely, as in the proof of Lemma 5.8, we systematically replace each occurrence of s' in the relations of $\mathcal{R}'$ with w_1. The diagram for $W/N(s'w_1)$ is then obtained by examining the effect of these replacements on the presentation.

The following lemma shows that the modifications necessary to create the new presentation from the original one are not substantial.

Lemma 5.10   *Each of the relations $(s')^2$, $(s'v')^2$ $(v' \in S_1')$, and $(s't')^n$ follows from the relations in the original presentation. If $a' \in S' \setminus S_1'$ and $(s'a')^m$ is a relation in the original presentation, then $m = 2$, and in adding the relation $(w_1a')^m$ to the presentation, we may also replace some relation $(v'a')^k$ $(v' \in S_1'$ and $k > 2)$ with $(v'a')^2$.*

Proof.   The first sentence is clear. Consider then a relation $(s'a')^m = 1$, $a' \in S' \setminus S_1'$. If $m > 2$, for $\sigma' \in \mathcal{M}(S')$ containing a' and s', $t' \notin \sigma'$. If $\sigma \in \mathcal{M}(S)$ is conjugate to σ', the minimality of the choice of τ_s forces $s \notin \sigma$. For each letter $b \in \sigma \cap \tau_s$, W_{τ_b} is abelian, so by the first case there is a letter $b \in \sigma \cap \tau'$ conjugate to b. It follows that some letter in $\sigma \cap \tau_s$ is conjugate to s'.

If s' and v were conjugate (for $v \in S_1$), then s' and v' would be conjugate, for some $v' \in S_1'$, which cannot be. Thus s' and t are conjugate. But then, since t is conjugate to $t'w_2$, we may obtain a contradiction by applying the quotient which identifies each of t and $v' \in S_1'$ to 1. Thus $m = 2$.

Now suppose that $a' \in (S' \setminus \tau') \cap \mathrm{St}(s')$. An argument much like that just given above (supply it!) shows that $[a't']$ is an edge with label 2.

This implies that if $a' \in (S' \setminus \tau') \cap \mathrm{St}(s')$, there exists $\sigma' \in \mathcal{M}(S')$ such that $\{a', s', t'\} \subseteq \sigma'$. If $\sigma \in \mathcal{M}(S)$ is conjugate to σ', then $s \in \sigma$ (why?), so that $\tau' \subseteq \sigma'$. Finiteness of $W_{\sigma'}$ forces there to be at most one edge $[a'v']$ $(v' \in S_1')$ with label exceeding 2. If there is no such edge, the relation $(w_1a')^2 = 1$ is a consequence of the original relations. If there is a single such edge, say $[a'v']$, then the relation $(w_1a')^2 = 1$ follows from the original relations and the additional relation $(v'a')^2 = 1$.   □

Note that the proof above shows precisely *which* edges must be relabeled in passing to the new diagram (namely, the edges $[a'v']$ for a' adjacent to s' and v' appearing in w_1).

Lemma 5.10 now allows us to compute a second diagram for $W/N(s) = W/N(s'w_1)$. However, this diagram cannot possibly correspond to the same group as the diagram obtained from $\mathcal{V}$ by removing s and all adjacent edges.

First note Lemma 5.10 shows that $[s't']$ is the only edge in $\mathcal{V}'$ which contains s' and has label greater than 2. Furthermore, by the definition of τ_s is it easy to see that $[st]$ is the only edge in $\mathcal{V}$ containing s and having label greater than 2. We claim now that the number of edges incident s in $\mathcal{V}$ is the same as the number of edges incident s' in $\mathcal{V}'$.

Indeed, examine the diagrams $\bar{\mathcal{V}}$ and $\bar{\mathcal{V}}'$ obtained from $\mathcal{V}$ and $\mathcal{V}'$, respectively, by removing the vertices S_1 from $\mathcal{V}$ and S_1' from $\mathcal{V}'$. These diagrams correspond to the same group, since $W/N(S_1) = W/N(S_1')$. Therefore, by Theorem 5.2, $\bar{\mathcal{V}}$ and $\bar{\mathcal{V}}'$ have the same number of edges. We further modify these two diagrams by removing the vertices corresponding to the $N(S_1)$-coset of s (these are none other than the vertices s in $\mathcal{V}$ and s' in $\mathcal{V}'$. Since the resulting diagrams both correspond to $W/N(S_1 \cup \{s\}) = W/N(S_1' \cup \{s'\})$, they must have the same number of edges.

It follows that the number of edges incident s is the same as the number of edges incident s'.

With the following lemma we will at last be able to obtain a contradiction:

Lemma 5.11 *Let v be a letter (in S_1) appearing in w_1, and let v' be the letter of S_1' conjugate to it. Then there is a letter $a' \in S'$ such that a' is adjacent to both s' and v' and $[a'v']$ has label greater than 2.*

Proof. Because $\mathrm{St}(x) \not\subseteq \mathrm{St}_2(v)$ for every v appearing in w_1, there is some letter $a \in S$ adjacent to both s and v such that $[av]$ is an edge with label exceeding 2. Choosing $\sigma \in \mathcal{M}(S)$ containing $\{a, s, v\}$ and $\sigma' \in \mathcal{M}(S')$ conjugate to σ, it is not difficult to show that v' is not central in the group $W_{\sigma'}$. (Otherwise v would be central in W_σ, contradicting the choice of σ.) Thus, since $\tau' \subseteq \sigma'$ (why?), σ' contains some letter a' which possesses the properties we seek. $\qquad \square$

Why is Lemma 5.11 enough to complete the proof of Theorem 3.8? In removing s and all incident edges from $\mathcal{V}$, we obtain a diagram with one fewer edge with label greater than 2. In removing s' from $\mathcal{V}'$ and relabeling edges as instructed to by Lemma 5.10, at least 2 edges with label exceeing 2 are lost (unless $w_1 = 1$). This contradicts Theorem 5.2. Therefore w_1 must be trivial after all, and s' is indeed conjugate to some generator in S'.

We are now in a position to prove Theorem 3.4, which we restate here with slightly different notation:

Corollary 5.3 *Let (W, S) be a right-angled Coxeter system with diagram*

$\mathcal{V}$. Then (W, S) *is strongly reflection rigid if and only if the following condition holds:*

1. *for every vertex s in $\mathcal{V}$ the full subgraph on the set of vertices not in* St(s) *is connected.*

   Moreover, W is strongly rigid if and only if the following condition also holds:

2. *for every vertex $s \in \mathcal{V}$, $\tau_s = \{s\}$.*

Proof. Let (W, S) be a right-angled system, and suppose that the first condition above fails for the vertex s. Choose a connected component U of the full subgraph on $S \setminus$ St(s), and define a map α by $\alpha(u) = sus$ for every $u \in U$, and $\alpha(u) = u$ for $u \in S \setminus U$. It is easily checked that this is an automorphism which cannot be inner, yet the reflections relative to the newly defined generating set coincide with those of (W, S). Thus if such a vertex s exists, (W, S) is not strongly reflection rigid. The converse is proven by an application of Theorem 7.2, as the reader should verify.

The second statement ("Moreover...") now follows at once from Theorem 3.8. $\square$

5.5   The Isomorphism Problem for even Coxeter groups

5.5.1   *The algorithm*

In the case of even Coxeter groups, we have as good a solution to the question of isomorphism as we could ever want:

Theorem 5.8   *There exists an algorithm which determines whether or not the group W defined by an arbitrary Coxeter system (W, S) is an even Coxeter group (with respect to any system). Moreover, the algorithm describes how to construct this even system (and diagram) if such a system exists.*

That this gives a solution to the Isomorphism Problem in the case of even Coxeter groups is clear. Begin with (W, S) and (W', S') and apply the algorithm mentioned above. If only one of the groups W and W' is even, they cannot be isomorphic. If both are, Theorem 5.4 allows us to check whether $W \cong W'$ by merely comparing the (unique) even diagrams computed by applying the algorithm of Theorem 5.8.

As indicated by Theorem 3.10, the algorithm we seek involves successive applications of diagram twists and edge-triangle exchanges in which a triangle with edge label multiset $\{2, 2, k\}$ (k odd) is replaced by an edge labeled $2k$.

In case $\mathcal{V}$ is an even diagram, all diagram twists are trivial twists. Even if $\mathcal{V}'$ is a non-even diagram corresponding to an even group, the sort of twists that can be performed is quite limited. Namely, we will only need to consider twists across a single edge $[s't']$ with an odd label. Such a twist consists of replacing each edge $[x's']$ with an edge $[x't']$ (with the same label), and *vice versa*, switching the roles of s' and t'.

Recognizing when twists can and need to be performed is key. The following proposition forms the backbone of our algorithm:

Proposition 5.10 *Suppose that $\mathcal{V}'$ is a (non-even) diagram corresponding to the even group W.*

1. *For every edge $[s't']$ of $\mathcal{V}'$ with an odd label, both s' and t' have valence at least 2. (In particular, $[s't']$ must lie on some simple circuit of $\mathcal{V}'$.)*
2. *If $\mathcal{V}'$ contains an edge $[s'_1 s'_2]$ with odd label which lies on a simple circuit $\{[s'_1 s'_2], [s'_2 s'_3], ..., [s'_k s'_1]\}$, then either this circuit has length 3 or there exists a chord $[s'_i s'_j]$ ($|i-j| \geq 2$) which bisects the circuit into two shorter circuits.*
3. *Given any edge $[s't']$ in $\mathcal{V}'$ with odd label, we can obtain a diagram $\mathcal{V}''$ from $\mathcal{V}'$ by first performing a (perhaps trivial) twist across the edge $[s't']$ and then replacing some triangle containing $[s't']$ by an edge with an even label.*

Given a non-even diagram $\mathcal{V}'$, we use Proposition 5.10 to attempt to reduce the number of odd edges which it contains; if we cannot perform this reduction, the group W corresponding to $\mathcal{V}'$ cannot be even.

Though it is very elementary and not theoretically difficult, the proof of Proposition 5.10 is highly technical and involved. We will prove only the first and third statements, assuming the second is true (its proof is the most technical of the three). The foundation of the entire proof is an understanding of the combinatorics of the non-even diagram $\mathcal{V}$ corresponding to the even group W.

5.5.2 *Matching edges between even and non-even diagrams*

Until further notice we suppose that $\mathcal{V}$ is an even diagram, while $\mathcal{V}'$ and $\mathcal{V}''$ are arbitrary diagrams, all corresponding to the same even group W. We denote by (W, S), (W, S'), and (W, S'') the corresponding systems. The reader is strongly encouraged to not be shy in drawing pictures while reading the following pages, as the correspondences between these diagrams can become quite confusing!

Our first goal is to understand how the edges of each of these diagrams correspond to those of another. The following result is of fundamental importance in this regard:

Proposition 5.11 *Let V' and V'' be as above. Then there is a unique bijection ϕ between the edges of V'' with label greater than 2 and the edges of V' with label greater than 2 such that if $\phi([s''t'']) = [s't']$, then exactly one of the following conditions obtains (for some $k \geq 1$ and some p satisfying $(p, k) = 1$):*

1. *$[s't']$ and $[s''t'']$ are both labeled $2k + 1$, $s''t'' \sim (s't')^p$, and $\langle s''t'' \rangle \sim \langle s't' \rangle$,*
2. *$[s't']$ is labeled $2(2k + 1)$, $[s''t'']$ is labeled $2k + 1$, $s''t'' \sim (s't's't')^p$, and $\langle s''t'' \rangle \sim \langle s't's't' \rangle$,*
2'. *2. holds, with the roles of V' and V'' reversed,*
3. *$[s't']$ and $[s''t'']$ are both labeled $2(2k + 1)$, $s''t''s''t'' \sim (s't's't')^p$, and $\langle s''t''s''t'' \rangle \sim \langle s't's't' \rangle$, or*
4. *$[s't']$ and $[s''t'']$ are both labeled $4k$, $s''t''s''t'' \sim (s't's't')^p$, and $\langle s''t''s''t'' \rangle \sim \langle s't's't' \rangle$.*

Proof. As in [Mihalik (preprint 1)], we make an initial reduction by assuming that one of the diagrams (say, V') is even, effectively eliminating the first case from consideration. (Why can this be done?) Now take an edge $[s''t'']$ in V'' with label greater than 2.

Let $\sigma' \subseteq V'$ and $\sigma'' \subseteq V''$ be maximal spherical simplices such that $[s''t''] \subseteq \sigma''$ and $W_{\sigma'} \sim W_{\sigma''}$. Conjugating, we may assume that $W_{\sigma'} = W_{\sigma''}$. On the one hand, this group decomposes as a direct product of dihedral groups and copies of $\mathbb{Z}_2$, as usual, with one dihedral factor corresponding to $[s''t'']$. (More precisely, there is a dihedral factor corresponding to each edge of σ'' with label exceeding 2; the order of this group is twice the edge label whenever that label is odd or divisible by 4, and it is equal to the edge label whenever the label is of the form $2(2k + 1)$, $k \geq 1$.)

On the other hand, we obtain a similar direct product decomposition involving edges of V'. Theorem 5.1 gives a one-to-one correspondence between the factors in these two decompositions. The following lemma tells us how to define ϕ:

Lemma 5.12 *Let the group G decompose as a direct product in two dif-*

ferent ways, $\prod_{i=1}^{r} A_i$ *and* $\prod_{i=1}^{r} B_i$, *where each* A_i *and each* B_i *is either* $\mathbb{Z}_2$ *or* D_k *for* k *not of the form* $2(2k+1)$. *Then for each dihedral factor* $B_j = \langle b_1, b_2 \rangle$ *there is a unique integer* i *such that* $A_i = \langle a_1, a_2 \rangle$ *has the same order as* B_j *and*

1. *if* $|B_j|$ *is* $2n$ *for odd* n, *then* $b_1 b_2 = (a_1 a_2)^p$ *for some* p *satisfying* $(p, n) = 1$, *and* $\langle b_1 b_2 \rangle = \langle a_1 a_2 \rangle$, *and*
2. *if* $|B_j|$ *is* $2n$ *for even* n, *then* $b_1 b_2 = (a_1 a_2)^p \alpha$ *for some* p *satisfying* $(p, n) = 1$ *and some* α *whose order divides* 2 *and which commutes with* a_1 *and* a_2. *In this case* $\langle b_1 b_2 b_1 b_2 \rangle = \langle a_1 a_2 a_1 a_2 \rangle$.

The proof of this fact is elementary and is left to the reader as an exercise. Lemma 5.12 gives us an obvious means of associating to each edge of σ'' with label greater than 2, a similar edge in σ'. Because $\mathcal{V}'$ is even, any two of its edges lie in different conjugacy classes. (This is proven most easily using quotient arguments.) Therefore the matching induced by Lemma 5.12 is well-defined, even when we extend it to a map from all edges of $\mathcal{V}''$ with labels greater than 2. This extended function clearly maps onto the set of edges of $\mathcal{V}'$ with label greater than 2. Once we show that it is one-to-one, Proposition 5.11 will be proven, since all of the stated combinatorial properties are inherited from Lemma 5.12.

Throughout the remainder of this chapter, ϕ denotes the edge matching defined above. The next lemma allows us to speak about triangles (and circuits) containing odd edges:

Lemma 5.13 *Every edge of* $\mathcal{V}''$ *with an odd label is contained in some triangle of* $\mathcal{V}''$.

Note that this implies the first part of Proposition 5.10.

Proof. Denote by τ'' the intersection of all simplices of $\mathcal{V}''$ which contain $[s''t'']$ and which generate a subgroup conjugate to some parabolic subgroup of (W, S'). Then τ'' contains at least $[s''t'']$. However, since $W_{\{s'', t''\}}$ is not an even Coxeter group, Proposition 4.1 shows that there must be more to τ'' than $[s''t'']$, so there is a vertex $x'' \in \mathcal{V}''$ such that

$$x'' \in \mathrm{Lk}(s'') \cap \mathrm{Lk}(t'').$$ □

Lemma 5.14 *For any triangle* $\{[a''b''], [b''c''], [c''a'']\}$ *in* $\mathcal{V}''$ *containing an odd edge* $[a''b'']$, *one of the following holds:*

1. *all edges of the triangle have label greater than 2 and* $\phi([a''b'']) = \phi([b''c'']) = \phi([c''a''])$,

2. *two edges (say* $[a''b'']$ *and* $[b''c'']$*) have label greater than* 2, *and* $\phi([a''b'']) = \phi([b''c''])$, *or*

3. $[a''b'']$ *is the only edge of the triangle with label greater than* 2.

Proof. Recall that $\mathcal{V}'$ is assumed to be even.

Let $[a''b'']$ have odd label $2k + 1$, so that $[a'b'] = \phi([a''b''])$ has label $2(2k + 1)$. Suppose also that $[a''c'']$ has label $m > 2$, and that $[a_1'c'] = \phi([a''c'']) \neq [a'b']$, making no assumptions on $[b''c'']$'s label yet.

We first consider the quotient map $\nu : W \rightarrow W/N(a''b'') = W/N(a'b'a'b')$. (The equality here follows from properties of ϕ already established.) By considering the even diagram $\mathcal{V}'$, we see that ν is injective on $W_{\{a_1',c'\}}$. Suppose now that $[b''c'']$ has label 2. Then

$$\nu(a''c''a''c'') = \nu(a''b''c''b''a''b''c''b'') = (\nu(c'')\nu(b''))^2 = 1 \qquad (5.6)$$

so that $(a''c'')^2$ is in the kernel of u.

There are two possibilities, from the definition of ϕ: one of $\langle a''c'' \rangle$ or $\langle (a''c'')^2 \rangle$ is conjugate to $\langle (a_1'c')^2 \rangle$. (In the first case, $(a_1'c')^2$ must have odd order at least 3, so the label on $[a_1'c']$ is at least 6.) However, applying (5.6) gives a contradiction to the injectivity of ν on $W_{\{a_1',c'\}}$. So $[b''c'']$ must have label at least 3.

Now assume $\phi([b''c'']) = [a'b']$. In this case the label on $[b''c'']$ is either $2k+1$ or $2(2k+1)$. In the first instance, $N(b''c'') = N(a'b'a'b') = N(a''b'')$, so $a''c'' \in N(b''c'')$, and from this $a_1'c'a_1'c' \in N(a''b''a''b'')$, which cannot be the case, since $\mathcal{V}'$ is even. The other possibility yields a similar contradiction. Furthermore, very similar computations give contradictions in case $\phi([b''c'']) = [a_1'c']$.

Suppose instead that $\phi([b''c'']) = [b_1'c_1']$ for some third edge $[b_1'c_1'] \subseteq \mathcal{V}'$. Note that $c''a'' = c''b''(a''b'')^{2k}$, so $\nu(c''a'') = \nu(c''b'')$. If both $[a''c'']$ and $[b''c'']$ have odd labels, then $a''c'' \sim a_1'c'a_1'c'$ and $b''c'' \sim b_1'c_1'b_1'c_1'$. Applying ν all around we obtain $\nu(\langle a_1'c'a_1'c' \rangle) = \nu(\langle b_1'c_1'b_1'c_1' \rangle)$, which is nonsense, since $\mathcal{V}'$ is even. Similar problems arise for other choices of the parities of $[a''c'']$ and $[b''c'']$.

Therefore $\phi([a''c'']) = [a'b']$ must hold, as desired, so long as any edge of the triangle $\{a'', b'', c''\}$ has label greater than 2. $\square$

Now we refine the previous two lemmas simultaneously:

Lemma 5.15 *For every edge* $[a''b'']$ *of* $\mathcal{V}''$ *with odd label, any triangle containing* $[a''b'']$ *has two edges labeled* 2.

From this it follows immediately that ϕ is injective. Indeed, for any edge $[s''t'']$ of $\mathcal{V}''$ with an odd label, Lemma 5.15 shows that a diagram for the group $W/N(s''t'')$ can be found merely by identifying the vertices s'' and t''. (To prove this, consider presentations for these groups.) Therefore distinct edges of $\mathcal{V}''$ cannot correspond under ϕ to the same edge of $\mathcal{V}'$: only one of them factors through the quotient $\nu : W \to W/N(s''t'')$.

Proof.   Were Lemma 5.15 false, we could find a counterexample $\mathcal{V}''$ which is minimal in that the product of its edge labels is as small as possible. Let $\mathcal{V}''$ be such a diagram, and let $\mathcal{V}'$ be the unique even diagram corresponding to the same system.

By Lemma 5.14, $\mathcal{V}''$ contains a triangle with vertices $\{a'', b'', c''\}$ in which at least two odd-labeled edges correspond to $\phi([a''b'']) = [a'b']$. Let $\sigma'' \subseteq \mathcal{V}''$ be a maximal simplex containing the triangle $\{a'', b'', c''\}$. (Be careful: σ'' is a maximal simplex, not a maximal *spherical* simplex!)

We rely on the following fact, proven by M. Mihalik in [Mihalik (preprint 2)]:

Proposition 5.12   *Given any two Coxeter systems (W, S) and (W, S') for the same group W, there is a one-to-one correspondence between the conjugacy classes of parabolic subgroups generated by maximal simplices in the respective diagrams $\mathcal{V}$ and $\mathcal{V}'$.*

This is shown by proving that the subgroups generated by maximal simplices are precisely the maximal FA subgroups, where the property FA is defined in [Serre (1980)].

Applying Proposition 5.12, we obtain a maximal simplex σ' in $\mathcal{V}'$ such that $W_{\sigma'}$ and $W_{\sigma''}$ are conjugate. Because $\mathcal{V}'$ is even, our minimality assumption allows us to conclude that $\mathcal{V}'' = \sigma''$ and $\mathcal{V}' = \sigma'$ (why?).

Let $\tau'' \subseteq \sigma''$ comprise the vertices of σ'' which are incident an edge e'' of σ'' for which $\phi(e'') = [a'b']$. If $s'' \in \tau''$ and $t'' \notin \tau''$, then Lemma 5.14 implies that $[s''t'']$ has label 2, so W decomposes as a direct product: $W \cong W_{\tau''} \times W_{\sigma'' \backslash \tau''}$.

If some edge $[s''t'']$ of $\mathcal{V}''$ had an even label greater than 2, we could find a smaller counterexample by considering the quotient $W/N(s''t''s''t'')$, which yields a diagram in which $[s''t'']$ has label 2. If there were an edge $[s''t'']$ with an odd label and $s'', t'' \in \sigma'' \backslash \tau''$, the quotient map $\nu : W \to W/N(s''t'')$ is injective on $W_{\tau''}$, so we could again find a smaller counterexample (by squashing out the edge $[s''t'']$).

Thus the only edge of $\mathcal{V}'$ which does not have label 2 is $[a'b']$, so that W is a finite direct product of dihedral groups, which cannot contain a triangle subgroup with more than one odd-labeled edge (*cf.* Lemma 4.2). $\square$

Applying the remark preceding the proof of Lemma 5.15, we have proven Proposition 5.11. $\square$

We state without proof further properties of the edge matching ϕ which will be needed in the following subsection. These facts can be verified by considering minimal counterexamples in order to derive a contradiction, and by judicious application of the appropriate quotient maps.

Proposition 5.13 *Let ϕ be the edge matching defined in Proposition 5.11, and assume that $\mathcal{V}'$ is even.*

1. *If $\phi([x''y'']) = [x'y']$ and $\phi([y''z'']) = [y_1'z']$ for distinct edges $[x''y'']$ and $[y''z'']$ of $\mathcal{V}'$, then $|\{x',y'\} \cap \{y_1',z'\}| = 1$, so the corresponding edges $[x'y']$ and $[y_1'z']$ share exactly one vertex.*
2. *If $\phi([x''y'']) = [x'y']$ and $\phi([y_1''z'']) = [y'z']$ for distinct edges $[x'y']$ and $[y'z']$ of $\mathcal{V}'$, either $|\{x'',y''\} \cap \{y_1'',z''\}| = 1$ or there is a path from some vertex of $\{x'',y''\}$ to some vertex of $\{y_1'',z''\}$, each of whose edges has an odd label.*
3. *If $x',y',z',x'',y'',y_1'',z''$ are as in 2. and $\{x',y',z'\}$ generates a triangle in $\mathcal{V}'$, then $y'' = y_1''$ and $\{x'',y'',z''\}$ generates a triangle in $\mathcal{V}''$.*
4. *If $[x''y''] \subseteq \mathcal{V}''$ has an odd label and $\phi([x''y'']) = [x'y']$, then one of the vertices x' and y' is incident no edge with label greater than 2 except for $[x'y']$.*

Assume for the moment that we have proven the second statement of Proposition 5.10. (That is, every simple circuit of length at least 4 containing the odd edge $[s't']$ has a "shortcut" chord.) In this case, every edge path connecting s' to t' either contains the edge $[s't']$ or intersects $\mathrm{Lk}_2(s') \cap \mathrm{Lk}_2(t')$. Denote by U the union of all connected components C of $\mathcal{V}' \setminus [\{s',t'\} \cup (\mathrm{Lk}_2(s') \cap \mathrm{Lk}_2(t'))]$ such that there is an edge $[s'x']$ for some $x' \in C$. For any edge $[x'y']$ for $x' \in U$ and $y' \in \mathcal{V}' \setminus (\{s',t'\} \cup U)$, the definition of U implies that $y' \in \mathrm{Lk}_2(s') \cap \mathrm{Lk}_2(t')$. Thus we can apply a diagram twist to the set U (twisting across $[s't']$), obtaining a new diagram, $\mathcal{V}''$, in which every edge incident s' (except $[s't']$) has label 2.

We are almost ready to prove the final statement of Proposition 5.10. We claim that we can find $u'' \in \mathrm{Lk}(s')$ such that

(U1) $u'' \in \mathrm{Lk}_2(s') \cap \mathrm{Lk}_2(t')$,
(U2) for every edge $[u''v'']$ in $\mathcal{V}''$, $v'' \in \mathrm{Lk}_2(s') \cap \mathrm{Lk}_2(t')$, and

(U3) for all simplices $\sigma'' \subseteq \mathcal{V}''$ containing u'' (respectively, containing s' and t') such that $W_{\sigma''} \sim W_\sigma$ for some $\sigma \subseteq \mathcal{V}$, s' and t' (respectively, u'') lie in σ''.

Once we find such a vertex u'', we may replace the triangle in $\mathcal{V}''$ generated by $\{s', t', u''\}$ (with edge labels $\{2, 2, 2k+1\}$) with a single edge with label $2(2k+1)$. Verifying the details of this replacement (that is, figuring out which vertices are connected to which in the new diagram) is left to the reader. Proposition 5.10's final assertion will follow.

5.5.3 *Finding u''*

Throughout the remainder of this section, s', t', and u'' retain their meanings from the previous subsection. We assume the necessary diagram twisting has been performed, and replace $\mathcal{V}'$ with $\mathcal{V}''$ if need be, in order to avoid excessive "primes". Given any simplex σ in any diagram, we denote by $|\sigma|$ the number of vertices in σ.

We already know (Lemma 5.13) that we can find a vertex u_1 in $\mathcal{V}'$ satisfying Condition (U1) from the previous section. We can easily improve this:

Lemma 5.16 *There exists a vertex $u_{1,3} \in \mathcal{V}'$ satisfying Conditions (U1) and (U3).*

Proof. We invoke the now familiar minimality assumption, letting the product of the edge labels of $\mathcal{V}'$ be as small as possible among all counterexamples. Let τ' be the intersection of all simplices σ' of $\mathcal{V}'$ containing $[s't']$ for which there is a simplex $\sigma \subseteq \mathcal{V}$ satisfying $W_\sigma \sim W_{\sigma'}$. (Compare the proof of Lemma 5.13; as in that lemma, $\tau' \neq \{s', t'\}$.) For any vertex $x' \in \tau' \setminus \{s', t'\}$, there can be no path from x' to either s' or t', by our restrictions on triangles.

Assume that for each $x' \in \tau' \setminus \{s', t'\}$, assume that there are simplices $\sigma'_{x'} \subseteq \mathcal{V}'$ and $\sigma_{x'} \subseteq \mathcal{V}$ such that $W_{\sigma_{x'}} \sim W_{\sigma'_{x'}}$, $x' \in \sigma'_{x'}$, and $\{s', t'\} \not\subseteq \sigma'_{x'}$. We can assume that $\sigma'_{x'} \subseteq \tau'$ for all x'. For any such x', we may find a smaller counterexample than $\mathcal{V}'$ in the following fashion. Let $z : W_{\sigma'_{x'}} \to \mathbb{Z}_2$ (written multiplicatively as $\mathbb{Z}_2 = \{-1, 1\}$) be the map which takes each element of $\sigma'_{x'} \setminus \{s', t'\}$ to 1 and $\sigma'_{x'} \cap \{s', t'\}$ to -1. Then $W/N(\ker(z))$ yields the desired counterexample. (Prove this by computing diagrams for the latter group.)

Therefore some vertex $x' \in \tau' \setminus \{s', t'\}$ satisfies (U3), as desired, giving us our $u_{1,3}$.   □

In order to finish the proof, we need only show that $u_{1,3}$ can be chosen so that every edge incident this vertex has label 2. If this cannot always be done, we may assume that $\mathcal{V}'$ a minimal counterexample, exactly as before.

Let τ' be defined as in the proof of Lemma 5.16. By hypothesis, every vertex in τ' satisfying (U3) is incident some edge with label exceeding 2.

We may use the minimality assumption to show that every edge of $\mathcal{V}'$ whose label exceeds 2 must contain a vertex of τ' satisfying (U3). Indeed, let $[x'y']$ be an edge with label $m > 2$, and assume neither vertex satisfies (U3). By applying the appropriate quotient homomorphism (mapping to the group $W/N(x'y'x'y')$ if m is even, and to $W/N(x'y')$ if m is odd), we may produce a new group (with corresponding diagrams) which still gives a contradiction and which is smaller in our edge-label measure than is $\mathcal{V}'$.

Let $\tau \subseteq \mathcal{V}$ be the simplex in $\mathcal{V}$ satisfying $W_\tau \sim W_{\tau'}$.

Lemma 5.17 $|\tau| < |\tau'|$.

Proof. By applying the appropriate quotient homomorphisms we can obtain from τ and from τ' diagrams corresponding to the same right-angled Coxeter group. By rigidity of such groups, the number of vertices remaining after applying the quotients must be identical. However, in modifying τ' we must collapse at least one odd-labeled edge by identifying its vertices. No such identification is performed on the even diagram τ, so we must have had strictly more vertices in τ' than in τ.   □

Denote by ω (ω') the subset of τ (τ') consisting of vertices which are incident some edge of $\mathcal{V}$ ($\mathcal{V}'$) with label exceeding 2.

Lemma 5.18 *The number of vertices of ω incident an edge of τ with label exceeding 2 is equal to the number of vertices of ω' incident an edge of τ' with label exceeding 2.*

Proof. This follows easily from Proposition 5.11.   □

Lemma 5.19 *Let x' be a vertex of τ' and let $[x'y']$ be an edge with label greater than 2, $y' \notin \tau'$. Let $[xy]$ be the edge of $\mathcal{V}$ corresponding to $[x'y']$. Then exactly one of x and y (say, x) lies in τ, and x does not lie on an edge of τ with label exceeding 2.*

Proof. Obviously our edge matching forces $\{x,y\} \not\subseteq \tau$; relabel if needed so that $x \in \tau$. (Note that one of the vertices x, y must lie in τ, which can be seen by forming the quotient $W/N(\tau)$.) Suppose that $[xz]$ is an edge in τ with label exceeding 2, and let $\phi([x_1'z']) = [xz]$ for $[x_1'z'] \subseteq \tau'$.

By Proposition 5.13 there is a path between the edges $[x'y']$ and $[x'_1 z']$ in which every edge has an odd label. The first edge of this path cannot be $[y'v']$ for any vertex v', for otherwise $v' \in \tau'$, and since τ' is a simplex, $\{x', y', v'\}$ would give a triangle contradicting Lemma 5.15.

Therefore the first edge in the path must be $[x'v']$ for some vertex $v' \notin \tau'$ (by our assumption on x'). Since $v' \notin \tau'$, there must be at least one more edge in the path to $[x'_1 z']$. However, this edge yields a contradiction to Lemma 5.15, as in the previous paragraph. We must conclude that no edge $[xz]$ as above exists. $\square$

Lemma 5.20 *Let $[x'_1 y'_1]$ and $[x'_2 y'_2]$ be edges of $\mathcal{V}'$ with labels exceeding 2. Suppose furthermore that $x'_i \in \tau'$, but neither x'_i belongs to an edge of τ' with label exceeding 2, and that $y'_i \notin \tau'$. Let the edges $[x_i y_i] \subseteq \mathcal{V}$ $(i = 1, 2)$ correspond to $[x'_i y'_i]$ as in Proposition 5.11. Then up to relabeling, $x_i \in \tau$ and $y_i \notin \tau$ for $i = 1, 2$, and neither x_i is incident an edge of τ with label exceeding 2. Moreover,*

1. *if all four vertices x'_i and y'_i are distinct, then all four vertices x_i and y_i are distinct.*
2. *if $x'_1 = x'_2$ and $y'_1 \neq y'_2$, then $x_1 = x_2$ and $y_1 \neq y_2$, and*
3. *if $y'_1 = y'_2$ and $x'_1 \neq x'_2$, then $y_1 = y_2$ and $x_1 \neq x_2$.*

Proof. We prove a single case mentioned above (the second listed above) and leave the reader to investigate the (similar) proofs of the remaining cases.

By Proposition 5.13, $x_1 = x_2$ holds (after relabeling, if necessary). Moreover, the only other possible configuration (besides the desired one) for τ would have $x_1 \notin \tau$ and $y_i \in \tau$ for $i = 1, 2$, with neither of these vertices incident an edge of τ with label greater than 2. Suppose this were to occur.

We apply Proposition 5.12 to conclude that $\{x_1, y_1, y_2\} \subseteq \sigma \subseteq \mathcal{V}$ and $\{x'_1, y'_1, y'_2\} \subseteq \sigma' \subseteq \mathcal{V}'$ for some maximal simplices σ and σ' such that $W_\sigma \sim W_{\sigma'}$. From this, there is an edge $[y'_1 y'_2]$, and none of the edges of the triangle $\{x'_1, y'_1, y'_2\}$ have odd labels. Moreover, there is no odd-labeled path from $[y'_1 y'_2]$ to τ', and no such path exists from y'_1 to y'_2.

We argue as in the proof of Lemma 5.16, letting $z : W_{\tau'} \to \mathbb{Z}_2$ be the map taking $\tau' \setminus \{x'_1\}$ to 1 and x'_1 to -1. Then a diagram for $W/N(\ker(z))$ can be obtained from $\mathcal{V}'$ by removing all vertices $\tau' \setminus \{x'_1\}$, as well as vertices connected to x'_1 by odd-labeled paths. A diagram for the same group can be obtained from $\mathcal{V}$ by removing some vertices of τ and identifying all vertices of τ which remain. In any case, only one of the edges $[x_1 y_1], [x_1 y_2]$ remains in the quotient, which cannot be since $\{x_1', y_1', y_2'\}$ factors through the quotient, leaving two distinct edges.   □

An immediate consequence of the above lemma is

Corollary 5.4 $|\omega| = |\omega'|$.

We have not really used the fact that $[s't'] \subseteq \tau'$ in the above arguments. Therefore, we may prove the following lemma in exactly the same way:

Lemma 5.21   *If the simplex $\sigma \subseteq \tau'$ satisfies $W_{\sigma'} \sim W_\sigma$ for some simplex $\sigma \subseteq \mathcal{V}$, then $|\sigma \cap \omega| = |\sigma' \cap \omega'|$.*

We are almost done. Now we need only prove the following fact:

Lemma 5.22 $|\tau \setminus \omega| \geq |\tau' \setminus \omega'|$.

From this we obtain the desired contradiction $|\tau| \geq |\tau'|$ immediately.

Proof.   The only vertices of τ and τ' not yet accounted for are those which are incident no edges with labels greater than 2. Take $x' \in \tau' \setminus \omega'$. Of course, by hypothesis x' lies in some simplex $\sigma_{x'}'$ for which there is $\sigma_{x'} \subseteq \mathcal{V}$ satisfying $W_{\sigma_{x'}} \sim W_{\sigma_{x'}'}$, $\{s', t'\} \not\subseteq \sigma_{x'}'$. By intersecting with τ' (or τ), we can assume that $\sigma_{x'}' \subseteq \tau'$ (and $\sigma_{x'} \subseteq \tau$). The discussion following the proof of Lemma 5.16 shows that $W_{\sigma_{x'}}$ is right-angled, and Lemma 5.21 implies that

$$|\sigma_{x'} \cap \omega| = |\sigma_{x'}' \cap \omega'| \text{ and } |\sigma_{x'} \cap (\tau \setminus \omega)| = |\sigma_{x'}' \cap (\tau' \setminus \omega')|. \qquad (5.7)$$

We claim that for right-angled $W_{\sigma_i} \sim W_{\sigma_i'}$, $\sigma_i \subseteq \tau$ and $\sigma_i' \subseteq \tau'$,

$$\left| \bigcap_{i=1}^n \sigma_i \cap (\tau \setminus \omega) \right| = \left| \bigcap_{i=1}^n \sigma_i' \cap (\tau' \setminus \omega') \right|. \qquad (5.8)$$

One inequality is easy: since there are no odd edges in $\mathcal{V}$, $W_{\sigma'} \sim W_\sigma$ for some $\sigma \subseteq \cap_i \sigma_1$, where $\sigma' = \cap_i \sigma_i'$. In the other direction, odd-labeled paths may complicate matters.

Take the case $n = 2$; the general case is entirely similar. $W_{\sigma_1 \cap \sigma_2} \sim W_{\bar\sigma_i'}$, some $\bar\sigma_i' \subseteq \sigma_i'$, for $i = 1, 2$. Given any vertex $v' \in \bar\sigma_1'$, there is an odd-labeled path from v' to some vertex of $\bar\sigma_2'$. But the vertices with which we are concerned (those that also lie in $\tau \setminus \omega$) are incident only edges with label 2, and so cannot lie on such paths.

Therefore

$$|\sigma_1 \cap \sigma_2 \cap (\tau \setminus \omega)| = |\bar\sigma_i' \cap (\tau' \setminus \omega')| = |\bar\sigma_1' \cap \bar\sigma_2' \cap (\tau' \setminus \omega')| \leq |\sigma_1' \cap \sigma_2' \cap (\tau' \setminus \omega')|,$$

as desired.

Since the sets $\sigma'_{x'} \cap (\tau' \setminus \omega')$ cover $\tau \setminus \omega'$ (for $x' \in \tau' \setminus \omega'$), it is not hard to see that $|\tau' \setminus \omega'| \leq |\tau \setminus \omega|$, as needed. $\qquad\qquad\qquad\qquad\square$

We have finally succeeded in proving the third statement in Proposition 5.10, yielding an effective solution to the Isomorphism Problem in even Coxeter groups.

5.6 Exercises

1. Prove Proposition 5.3. (*Hint:* consider the abelianization of W.)

2. Prove Proposition 5.4.

3. Prove Proposition 5.5.

4. Prove that if (G, X) is a reflection system and H is a reflection subgroup of G, then $\chi(H) \subseteq H$, and that $X \cap H = X \cap \chi(H)$. (See Section 5.3 for the appropriate definitions.)

5. With the definitions from Section 5.3, show that the map N_H defined in Proposition 5.7 is a reflection cocycle for the reflection subgroup H.

6. Prove that if if (W, S) and (W, S') are systems satisfying $R(S) = R(S')$, then there is a one-one correspondence ϕ between the irreducible factors of (W, S) and the irreducible factors of (W, S') such that $\phi(W_T) \cong W_T$ for all such factors W_T.

7. Fill in the details of the proof of Proposition 5.9.

8. Prove Lemma 5.12.

9. We outline a proof of the following theorem: the even Coxeter group W fails to be rigid if and only if, in its unique even diagram $\mathcal{V}$, there is an edge $[st]$ with label $2(2k + 1)$ such that s is incident exactly one edge (namely, $[st]$) with label exceeding 2, and $\mathrm{Lk}(s) \subseteq \mathrm{St}_2(t)$.

    a. Prove that if $\mathcal{V}$ contains an edge as described above, W is not rigid. (*Cf.* Proposition 5.9.)

Supposing the theorem false, let $\mathcal{V}$ be a minimal counterexample, as usual, and suppose that $\mathcal{V}'$ is a diagram for W containing an odd-labeled edge, $[s't']$, such that $\phi([s't']) = [st]$.

    b. Prove that $[st]$ is the only edge of $\mathcal{V}$ with label exceeding 2.

c.   Prove that if $\sigma \subseteq \mathcal{V}$ and $\sigma' \subseteq \mathcal{V}'$ are such that $W_\sigma \sim W_{\sigma'}$, then $|\{s,t\} \cap \sigma| = |\{s',t'\} \cap \sigma'|$.

d.   Prove that there can be no vertices u and v in $\mathcal{V}$ such that $[su]$ and $[tv]$ are edges, but $[sv]$ and $[tu]$ are not. (*Hint*: if this were to occur, consider maximal spherical simplices containing $[st]$ and $[su]$, and corresponding conjugate simplices in $\mathcal{V}'$. Apply part c. to the intersection of these simplices in $\mathcal{V}'$ to conclude that exactly one of $\{s',t'\}$ is in this intersection. Finally, apply the quotient map which identifies to 1 all vertices of $\mathcal{V}$ except those in $\{s,t,u,v\}$, and examine the conjugacy of the images of the various vertices under this map.)

10. Suppose that (W,S) is a right-angled Coxeter system, and suppose that $R' \subseteq R(S)$ is a set of reflections. Can you describe an elementary algorithm which computes the set S' appearing in Theorem 5.7, with respect to which $\langle R' \rangle$ is a Coxeter system?

Chapter 6

More general groups

In this chapter we focus on Coxeter groups whose diagrams are somewhat less restricted. We begin by considering large-type (also called skew-angled) Coxeter systems. Of course, these groups are all two-dimensional. We obtain a robust solution to the problem of rigidity in this setting. From there we more on to a different class of two-dimensional systems.

6.1 Large-type Coxeter systems

Recall that the system (W, S) is said to be *large-type* if $2 < m_{st} \leq \infty$ for every $s \neq t$ in S. Our aim in this section is to prove Theorem 3.13, which we restate here for completeness (refer to Chapter 3 for the terminology):

Theorem 6.1 *Let (W, S) be a large-type Coxeter system with diagram $\mathcal{V}$.*

1. (W, S) *is reflection rigid, up to diagram twisting.*
2. *If furthermore $\mathcal{V}$ does not have a spike with label $2(2k+1)$ for any $k \geq 1$, then W is rigid up to diagram twisting.*
3. *Let $|S| \geq 3$. If $\mathcal{V}$ is edge-connected, then W is strongly rigid.*

The proof of this theorem will require us to define a new object (called a chamber system) on which a given Coxeter group acts. The group's action on a chamber system gives rise to "reflections" and "root systems" in much the same way that reflections and root sytems arise from the action of W on a linear space (see 1.2.2). Rigidity and strong rigidity can be thought of as measures of uniqueness of certain nice sets of reflections, in a way described below. Therefore an intimate understanding of these reflections allows characterization of rigidity properties.

We will also make pivotal use of the decompositions of Theorem 2.7 to complete the proof of Theorem 6.1 in its most general form.

6.1.1   *The chamber system associated to a Coxeter system*

In this section we define chamber systems, and present a number of facts concerning these objects which will be needed in proving Theorem 6.1. Few proofs will be given; as the content of this subsection is very "building theoretic", many of the proofs can be found in [Brown (1989)] or [Ronan (1989)]. The reader is asked to supply some of the simpler proofs as exercises, and is encouraged to construct enough examples to develop an intuition for the definitions and assertions which follow.

Suppose that (W, S) is an arbitrary Coxeter system. The *chamber system* associated to (W, S) is a graph $Ch = Ch(W, S)$ whose set of vertices C is W itself, and for which there is an edge $\{u, v\}$ if and only if $u^{-1}v \in S$. Clearly $u^{-1}v \in S \Rightarrow u^{-1}v = v^{-1}u$, so that we may assume Ch is an undirected graph. The vertices C of Ch are called its *chambers*, and the edges P of Ch are called its *panels*.

Adjacency between chambers can be refined: we say that chambers u and v are *s-adjacent* if $u^{-1}v = s$; s is then called the *type* of the panel $\{u, v\}$. This definition gives us a natural mapping from the set P of all panels to S, by taking each panel to its type.

For example, consider the 3-generated group

$$W = \langle a, b, c \mid a^2, b^2, c^2, (ab)^3, (bc)^2 \rangle.$$

A portion of the chamber system $Ch(W, S)$ is shown in Figure 6.1.

It is clear that W acts (on the left) in a natural fashion on Ch, taking chambers to chambers and panels to panels. The action on the panels is even type-preserving, as the reader should prove.

Moreover, this group action shows that the reflections $R = R(S)$ of W are closely related to the chamber system Ch. Given a reflection $r \in R$, we let $P(r)$ denote the set of all panels left invariant by r (note that unless $r = 1$ no panel is *fixed* by r). In Figure 6.1, for instance, $P(b)$ is indicated by bold line segments. $C(r)$ is then defined to be the set of chambers whose panels lie in $P(r)$. Clearly u and ru are adjacent if and only if both belong to $C(r)$.

$P(r)$ can be thought of as a hyperplane in which r acts by reflection, and we will in fact refer to $C(r)$ as the *wall* associated to r. The following lemma makes the case more precisely:

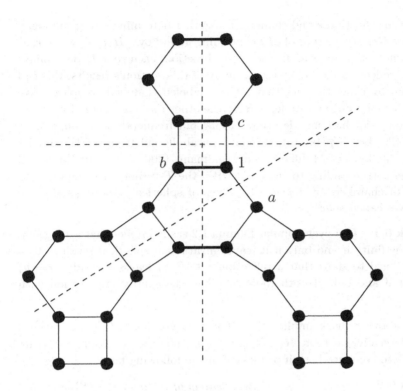

Fig. 6.1   A chamber system

Lemma 6.1   *Given any reflection r in the Coxeter system (W, S), the graph $Ch_r(W, S)$ obtained from $Ch(W, S)$ by removing every edge in $P(r)$ has precisely two connected components. Moreover, $\phi(u) = ru$ defines a one-to-one correspondence between the vertices of one component and the vertices of the other such that $\{u, v\}$ is an edge in the first component if and only if $\{\phi(u), \phi(v)\}$ is an edge in the second.*

We will not prove this lemma here; the most elementary proof follows by recognizing that Ch is what is what is known as a *labeled chamber complex* (*cf.* Section 1.5.2) as defined in [Brown (1989)]. Following [Brown (1989)] one can then derive a great deal more information about Ch.

We use Lemma 6.1 to define roots corresponding to the reflection r, much as roots were first defined in 1.2.2 relative to W's action on a given vector space. Now, a *root* is any one of the two components of a graph Ch_r, $r \in R(S)$. As in 1.2.2, we denote the set of roots by $\Phi = \Phi(W, S)$.

A given reflection r and chamber u together determine a root uniquely: denote by $H(r, u)$ the root of Ch_r containing u, and by $-H(r, u) = rH(r, u)$ the root not containing u. In the other direction, each root H determines a unique reflection $r \in R$. (As with many of this section's results, this can be proven by using the fact that Ch is a labeled chambed complex.) We denote this r by r_H, and we let $-H$ be the other root corresponding to r_H.

Not only do chambers determine roots, but frequently subgroups of W do as well. Let W_T be any standard parabolic subgroup of W and let $u \in W$. Following [Mühlherr and Weidmann (2002)] we call the set of chambers corresponding to the coset uW_T the *T-residue* of u, and denote this set of chambers by $R_T(u)$. Residues of spherical subgroups W_T are themselves called *spherical*.

Remark 6.1   Our ever-popular Lemma 1.2 easily implies that a subgroup G of W is finite if and only if it leaves invariant a spherical residue. It is also not hard to show that a reflection $r \in R(S)$ leaves a residue $R_T(u)$ invariant if and only if both roots of r have nonempty intersection with $R_T(u)$.

The above remark implies that if r does not leave $R_T(u)$ invariant, $R_T(u)$ determines a root, $H(r, R_T(u))$, of r. The reader should construct examples to convince herself or himself of the following fact:

Lemma 6.2   *Suppose G is a finite subgroup of W and $r \in R(S)$ such that $\langle G \cup \{r\} \rangle$ is an infinite subgroup of W. Then one of the roots associated to r contains every one of the spherical residues left invariant by G.*

Just as the wall $C(r)$ corresponds to the hyperplane in which r acts as a reflection, the roots associated to r behave much like the halfspaces into which a hyperplane divides a vector space. For instance, it is not hard to see that every root is convex (geodesics between two points of a single root lie entirely within that root), and therefore that the intersection of any two roots is connected.

6.1.2   *The chamber system $Ch(W, S)$ and reflection subgroups of W*

We continue our discussion of the chamber system $Ch = Ch(W, S)$ of (W, S) by indicating how it relates to subgroups of W generated by reflections.

Let $R' \subseteq R(S)$, and let $W(R')$ denote the subgroup generated by R'. Define two chambers u, v in Ch to be *R'-equivalent* (denoted $u \sim_{R'} v$) if for all $r \in R \cap W(R')$, $H(r, u) = H(r, v)$. The reflection $r \in R'$ is called an *R'-wall* for the chamber u if for some $v \in C(r)$, $H(r', u) = H(r', v)$ for all $r' \in R'$.

We return to the example given in the previous subsection, in which W is 3-generated. Let $R' = \{a, b, aba\}$, which generates the 6-element Coxeter subgroup $\{1, a, b, ab, ba, aba\} \le W$. Both a and b are R'-walls of the chamber 1, whereas aba is not, as the reader can easily check. It is clear that in this example there are 6 R'-equivalence classes, and that the intersection $H(a, 1) \cap H(b, 1)$ is a fundamental domain for the action of $W(R')$ on the chambers of $Ch(W, S)$.

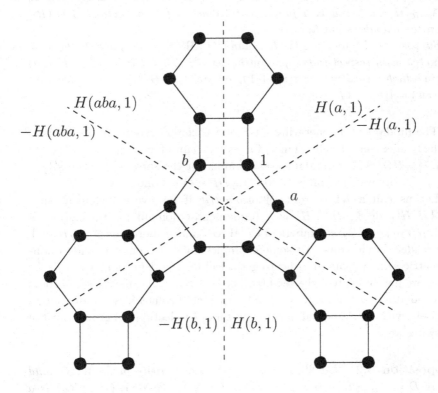

Fig. 6.2 *R'-walls and a fundamental domain for $W(R')$*

We obtain another instructive example by considering any Coxeter system (W, S) and letting $R' = R(S)$, so that $W(R') = W$. The only R'-walls of the chamber 1 are the fundamental generators, and each R'-

equivalence class consists of a single chamber. Moreover, the intersection $\cap_{s \in S} H(s, 1) = \{1\}$ is a fundamental chamber for the action of $W(R') = W$ on the chambers of $Ch(W, S)$.

These examples can be generalized completely:

Proposition 6.1 *Let $R' \subseteq R(S)$, and fix any chamber u in $Ch(W, S)$. Let R'_u denote the set of R'-walls of u.*

1. $(W(R'), R'_u)$ *is a Coxeter system.*
2. *The set of chambers D defined by $\{v \mid u \sim_{R'} v\}$ is also given by $\cap_{r \in R'_u} H(r, u)$, and is a fundamental domain for the action of $W(R')$ on the chambers of $Ch(W, S)$.*
3. *Suppose $r \in R'_u$ and $w \in W(R')$, and let $l : W(R') \to \mathbb{N}_0$ denote the word metric with respect to the generating set R'_u. Then either $wD \subseteq H(r, u)$ (in which case $l(rw) = l(w) + 1$), or $wD \subseteq -H(r, u)$ (in which case $l(rw) = l(w) - 1$).*

The fundamental generating set S of a Coxeter system (W, S) is a particularly nice set of reflections: for every pair of generators $s, s' \in S$, $H(s, 1) \cap H(s', 1)$ is a fundamental chamber for the subgroup $W(\{s, s'\}) = W_{\{s, s'\}}$. Moreover, the intersection $\cap_{s \in S} H(s, 1)$ is nonempty.

Let us call a set of roots Ψ *geometric* if for every 2-element subset $\{H, H'\}$ of Ψ, $H \cap H'$ is a fundamental domain for the action of $W(\{r_H, r_{H'}\})$ on the chambers of $Ch(W, S)$, and if moreover $\cap_{H \in \Psi} H \neq \emptyset$. The reader should construct some examples both of geometric and of nongeometric sets of roots in order to get a feel for the definition.

As we just indicated, the set $\{H(s, 1) \mid s \in S\}$ is a geometric set of roots corresponding to the system (W, S). In fact, geometric sets correspond precisely with fundamental generating sets of reflection subgroups, in the following way:

Proposition 6.2 *Let Ψ be a geometric set of roots in $Ch(W, S)$, and define $D = \cap_{H \in \Psi} H$ and $R_\Psi = \{r_H \mid H \in \Psi\}$. Then $(W(R_\Psi), R_\Psi)$ is a Coxeter system, and D is a fundamental domain for the action of $W(R_\Psi)$ on the chambers of $Ch(W, S)$.*

Exercise 5 asks the reader to develop some basic properties concerning geometric sets of roots.

6.1.3 *Universal root systems*

Let $R' \subseteq R(S)$ be a given set of reflections in (W, S). Following [Mühlherr and Weidmann (2002)] we call R' *universal* if $(W(R'), R')$ is a Coxeter system. For instance, S is obviously a universal set, as is any subset $T \subseteq S$.

The next proposition follows from M. Dyer's work on reflections in Coxeter groups (see Theorem 5.7 from the previous chapter, in particular):

Theorem 6.2 *Let (W, S) be an arbitrary Coxeter system, and let $R' \subseteq R(S)$ be a given finite set of reflections. Then there exists a universal set of reflections R'_0 such that $W(R'_0) = W(R')$.*

Thus in some sense there are many universal sets of reflections inside of $R(S)$. On the other hand, there is not a great deal of flexibility in choosing these sets. The first hint that this may be so is provided by the

Proposition 6.3 *Let (W, S) be an arbitrary Coxeter system. Suppose that $R' \subseteq R(S)$ is given such that R' is finite and for every two reflections $r_1, r_2 \in R'$, we may select corresponding roots H_i $(i = 1, 2)$ such that $\{H_1, H_2\}$ is geometric. Suppose that $W(R')$ is irreducible, and let Φ_1 and Φ_2 be two sets of roots such that*

$$R' = \{r_H \mid H \in \Phi_1\} = \{r_H \mid H \in \Phi_2\}$$

and every 2-element subset of Φ_i $(i = 1, 2)$ is geometric. Then

1. *if $\Phi_1 \cap \Phi_2 \neq \emptyset$, then $\Phi_1 = \Phi_2$, and*
2. *if $\Phi_1 \cap \Phi_2 = \emptyset$, then $\Phi_2 = \{-H \mid H \in \Phi_1\}$, and $r_1 r_2$ has finite order for every pair $r_1, r_2 \in R'$.*

Proof. Use Exercise 5. $\qquad\square$

In fact, we can say a bit more. Only finite reflection subgroups can correspond to distinct geometric sets of roots:

Proposition 6.4 *Let $R' \subseteq R(S)$ be a set of reflections as in Proposition 6.3. Let $\Phi_1 \neq \Phi_2$ be geometric sets of roots such that*

$$R' = \{r_H \mid H \in \Phi_1\} = \{r_H \mid H \in \Phi_2\}.$$

Then $W(R')$ is finite.

Proof. Because the fundamental domains associated with Φ_1 and Φ_2 are both nonempty, we may find a group element $w \in W(R')$ taking some chamber of the first domain to a chamber of the second. By the second part of Proposition 6.3, $\Phi_2 = -\Phi_2$, and this implies that the word w found above is the Garside element $\Delta_{W(R')}$ of $W(R')$ (see Exercise 1.7), implying that $W(R')$ is finite. □

Now we translate the above propositions into statements about universal sets of reflections, assuming for the moment that we are dealing only with strongly reflection rigid systems.

Proposition 6.5   *Let (W,S) be a Coxeter system with diagram $\mathcal{V}$. (W,S) is strongly reflection rigid if and only if for every universal set of reflections $R' \subseteq R(S)$ satisfying $\mathcal{V}_{(W(R'),R')} = \mathcal{V}$, we may select a geometric set of roots Φ such that $\Phi = \{\epsilon(r)H(r,1) \,|\, r \in R'\}$, for some $\epsilon : R' \to \{+,-\}$.*

Thus when (W,S) is infinite and strongly reflection rigid, every universal set of reflections with system isomorphic to (W,S) is essentially unique (by Propositions 6.3 and 6.4), and moreover this property characterizes strong reflection rigidity.

Proof.   Assuming strong reflection rigidity, the existence of the desired set of roots essentially follows from M. Dyer's work (*cf.* Theorem 6.2 above).

In the other direction, suppose (W,S) and (W,S') are systems for W with the same reflections. Let ν be an isomorphism of W taking each $s \in S$ to a reflection of (W,S'). Then ν takes the set S to a universal subset of the reflections $R(S')$, and by hypothesis $\nu(S)$ corresponds to a geometric set of roots Φ as desired. By definition, the intersection of the roots of Φ is a fundamental domain for the action of W on the set $Ch(W,S')$ of its chambers relative to S'. This intersection is thus a single chamber of $Ch(W,S')$, and $\nu(S)$ is therefore conjugate to the fundamental generating set S' of W'. □

6.1.4   *The first case: $\mathcal{V}$ is edge-connected*

In order to prove Theorem 6.1 in case $\mathcal{V}$ is edge-connected, we use the results of the previous section to detect the required rigidity.

Assume until futher notice that (W,S) is any Coxeter system with $R' \subseteq R(S)$ a universal set of reflections such that $(W(R'),R')$ is large-type and $\mathcal{V}_{R'}$ is edge-connected. Assume furthermore that $|R'| \geq 3$. If we can show that R' naturally corresponds to a geometric set of roots, then from the previous subsection we conclude that large-type, edge-connected Coxeter systems (generated by at least three elements) are strongly reflection rigid, nearly completing a proof of the third statement in Theorem 6.1.

Let us begin. We call a circuit $C = \{[s_1 s_2], ... [s_k s_1]\}$ in the diagram $\mathcal{V}$ *chord-free* if for any two vertices s_i and s_j which are not adjacent on C, there is no edge $[s_i s_j]$ in $\mathcal{V}$.

Lemma 6.3 *Let $T \subseteq R'$ correspond to a chord-free circuit C in the diagram $\mathcal{V}_{R'}$. Then T naturally corresponds to a unique geometric subset of roots.*

We denote the root H corresponding to $r \in T$ by $H(r, T)$.

Proof. Because $W(R')$ is large-type, such a chord free circuit generates an infinite parabolic subgroup of $W(R')$ which is easily seen to be strongly reflection rigid (see Exercise 7). Apply Proposition 6.5. $\square$

The idea now is simple: the hypothesis that $\mathcal{V}_{R'}$ is edge-connected implies that every vertex in this diagram lies on some chord-free circuit. The above lemma gives us a way of assigning to each element of R' some root. If we can show that the root assigned to each $r \in R'$ does not depend on the circuit containing r that we choose, we can construct a well-defined set of roots Φ which we can then show is geometric. Let us argue that this can be done by sketching the steps in the proof.

Remark 6.2 This argument is not unlike the one we will see below in Section 6.2, where, following [Bahls (2004a)], we adopt a combinatorial rather than a geometric point of view.

Proposition 6.6 *Let $r_1, r_2 \in R'$ such that $r_1 r_2$ has finite order, and let C and C' be chord-free circuits of $\mathcal{V}_{R'}$, both of which contain r_1 and r_2. Then $H(r_i, C) = H(r_i, C')$, $i = 1, 2$.*

Proof. If the circuits C and C' share at least three reflections in common (say r_3 in addition to r_1 and r_2), then we can "chain across" the subset $\{r_1, r_2, r_3\}$, which generates an infinite subgroup:

$$H(r_1, C) = H(r_1, \{r_1, r_2, r_3\}) = H(r_1, C').$$

(Prove this!) Otherwise, we use the fact that $\mathcal{V}_{R'}$ is edge-connected to construct a path outside of $C \cup C'$ linking a vertex of C to a vertex of C'.

More precisely, let $C = \{[r_1 t_C], ..., [r_2 r_1]\}$ and $C' = \{[r_1 t_{C'}], ..., [r_2 r_1]\}$, and let $P = \{[s_1 s_2], ..., [s_{n-1} s_n]\}$ be a path from $s_1 = t_C$ to $s_n = t_{C'}$ which includes neither r_1 nor r_2, as in Figure 6.3.

Assume in order to derive a contradiction that $H(r_1, C) = -H(r_1, C')$. From Exercise 5 we conclude that $H(r_2, C) = -H(r_2, C')$ as well. An easy

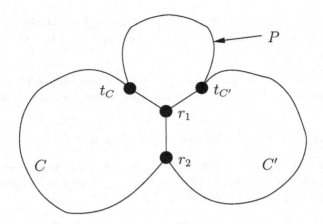

Fig. 6.3   An intermediate circuit

computation shows that

$$H(r_2, W(\{r_1, t_C\})) = H(r_2, C) = -H(r_2, C') = H(r_2, W(\{r_1, t_{C'}\})).$$

(See 6.1.1 for the definition of $H(r_2, G)$ for a group G.) Applying the action of r_1 we obtain

$$H(r_1 r_2 r_1, W(\{r_1, t_C\})) = H(r_1 r_2 r_1, W(\{r_1, t_{C'}\})).$$

Another easy computation now shows that

$$H(r_1 r_1 r_2, t_C) = -H(r_1 r_2 r_1, t_{C'}),$$

and a generalization of Exercise 4 completes the proof by yielding s_i in P such that $r_1 r_2 r_1 s_i$ has finite order, a contradiction.   □

What if two chord-free circuits share only a single vertex?

Proposition 6.7   *Suppose that r lies on both chord-free circuits C and C', and that $C \cap C' = \{r\}$. Then $H(r, C) = H(r, C')$.*

Proof.   We use a method that, suitably modified, will appear again in Section 6.2 below.

As in the previous proof, we add a path $\{[s_1 s_2], ..., [s_{n-1} s_n]\}$ linking vertices s_1 on C to s_n on C', such that $r s_1$ and $r s_2$ both have finite order. Now we step from C to the new circuit $\{[r s_1], P, [s_n r]\}$, and then to C', applying Proposition 6.6 at each step. (Note that we can choose P in order to ensure that the intermediate circuit is chord-free.)   □

Therefore we obtain a well-defined set of roots $\Phi = \Phi(R')$ by taking $H_r = H(r, C)$ for any chord-free circuit C containing $r \in R'$. Moreover, the above propositions show that every two-element subset $\{H_{r_1}, H_{r_2}\}$ is a geometric set of roots whenever $r_1 r_2$ has finite order. Thus in order to show that Φ itself is geometric, we must show that

1. $\{H_{r_1}, H_{r_2}\}$ is a fundamental domain for the action of $W(\{r_1, r_2\})$ whenever $r_1 r_2$ has infinite order, and
2. $\cap_{H \in \Phi} H \neq \emptyset$.

The first claim follows from computations very much like those in the proof of Proposition 6.6 (see [Mühlherr and Weidmann (2002)], Section 6, for details). To prove the second, we have the following

Proposition 6.8 *Let r_1, r_2 be distinct elements of R' such that $r_1 r_2$ has finite order. Let $R_T(u)$ be some spherical residue left invariant by $W(\{r_1, r_2\})$, and let $r \in R' \setminus \{r_1, r_2\}$. Then $R_T(u) \subseteq H_r$.*

Note that it then immediately follows that $R_T(u) \subseteq \cap_{H \in \Phi} H$, since $R_T(u) \cap H_{r_1} \cap H_{r_2} \neq \emptyset$ $(i = 1, 2)$ by Remark 6.1 (see Exercises 1 and 2). This same remark allows us to find the desired residue $R_T(u)$ in the first place.

Proof. First suppose that both $r_1 r$ and $r_2 r$ have finite order. Then we apply the proof of Proposition 6.6 to conclude that $H_r = H(r, W(\{r_1, r_2\}))$, which contains $R_T(u)$, by Exercise 2.

Otherwise, at least one of $r_1 r$, $r_2 r$ has infinite order. Say the former case holds. Similar computations show that $H_r = H(r, W(\{r_1\}))$. By Exercise 2 and the fact that $r_1 r$ has infinite order (so the walls corresponding to these reflections do not cross), $R_T(u)$ must lie in H_r, and we are done. $\square$

6.1.5 *The general case: applying visual decompositions*

At this point the reader may wish to review the definition of a *visual decomposition* given in Section 2.3 of Chapter 2. Exercises 2.9, 2.10, and 2.11 explore the structure of visual decompositions for different fundamental generating sets of W. The following result (which can be proven using Bass-Serre theory) takes these exercises one step farther:

Lemma 6.4 *Let (W, S) be a Coxeter system with spherical subgroup $C = W_T$ such that W does not decompose as a free product with amalgamation over any proper subgroup of C. Let $W = A_1 *_C \cdots *_C A_k$ be*

an amalgamated free product decomposition which is visual with respect to some fundamental generating set S' for W, and in which no A_i decomposes over C. (Such a decomposition exists, by Exercise 2.11.) Then for any fundamental generating set $R_1 \subseteq R(S)$ of reflections, there exists $R'_1 \subseteq R(S')$ which is twist equivalent to R_1 and with respect to which $A_1 *_C \cdots *_C A_k$ is a visual decomposition.

That is, given any set of reflections such that (W, R_1) is a Coxeter system, we may apply diagram twists to the diagram corresponding to R_1 to obtain a new diagram witnessing the desired visual decomposition of W. Let us now use this fact to establish Theorem 6.1.

Pick and fix the large-type system (W, S) with diagram $\mathcal{V}$. In order to test reflection rigidity, we let $\bar{S} \subseteq R(S)$ be some other fundamental generating set for W.

Remark 6.3 We show that there is an automorphism of W taking S to S', for some fundamental generating set $S' \subseteq R(S)$ which is twist equivalent to $\bar{S}$. It is left to the reader to translate this into a proof of reflection rigidity (up to twisting).

Suppose first that $\mathcal{V}$ is 0-connected but not edge-connected, and let $[st]$ be a cut-edge of $\mathcal{V}$, so that $W = W_1 *_C W_2$, for some parabolic subgroups W_i and $C = W_{\{s,t\}}$, $s, t \in S$, st of finite order n. By the hypothesis that $\mathcal{V}$ is 0-connected, W cannot decompose as a free product with amalgamation over any proper subgroup of $W_{\{s,t\}}$. In particular, $\mathcal{V}$ has no bridges.

Let $G = W_{\{s,t\}}$. We now use Exercise 2.11 to obtain a "finest" decomposition

$$W = A_1 *_G \cdots *_G A_k \tag{6.1}$$

which is visual with respect to some fundamental generating set S' twist equivalent to S. Lemma 6.4 allows us to replace $\bar{S}$ with $\bar{S}'$ to obtain a new fundamental generating set with respect to which (6.1) is a visual decomposition. Assume that we have made the replacements $S \mapsto S'$ and $\bar{S} \mapsto \bar{S}'$, if needed.

With these replacements, the decomposition $W_1 *_C W_2$ is visual with respect to both S and $\bar{S}$. Say $W_1 = \langle A \rangle = \langle \bar{A} \rangle$ and $W_2 = \langle B \rangle = \langle \bar{B} \rangle$. Conjugating, if needed, we may assume that $A \cap B = \bar{A} \cap \bar{B}$. (See Exercise 2.9.c.)

We may inductively assume that A and $\bar{A}$ are twist equivalent, and that B and $\bar{B}$ are as well. All twists that witness these equivalences involve conjugations which leave invariant the subset corresponding to the overlap, $A \cap B$. After performing these conjugations we obtain new sets (still denoted by A and B) such that A and $\bar{A}$ are conjugate, as are B and $\bar{B}$.

It is now an easy manner (using Corollary 2.1) to show that $S = \bar{A} \cup \bar{B}$ in fact holds, so that S and $\bar{S}$ are equal, and therefore the original generating sets are twist equivalent.

In case the diagram $\mathcal{V}$ contains a cut-vertex, s, but no bridge, the argument is only slightly more complicated: we must use Proposition 2.1 to prove the invariance of the overlap $A \cap B$ instead of the more straightforward Corollary 2.1.

What if $\mathcal{V}$ contains a bridge? By arguing inductively as in the last two cases, we obtain isomorphic subdiagrams corresponding to the conjugate sets $A \sim \bar{A}$ and $B \sim \bar{B}$. We can only obtain two non-isomorphic diagrams for S and $\bar{S}$ by pasting these subdiagrams together at different vertices as in Figure 6.4.

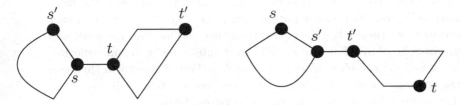

Fig. 6.4 $\mathcal{V}$ contains a bridge

However, each of the twists used in passing from A to $\bar{A}$ (or from B to $\bar{B}$) preserves conjugacy classes of reflections. Thus, the vertices s and s' in Figure 6.4 are conjugate to one another; likewise $t \sim t'$. Because two generators of a Coxeter system are conjugate to one another if and only if there is an odd path between them, we can perform successive twists along the edges of the odd paths connecting s to s' and t to t', obtaining at last our desired twist equivalence.

Remark 6.4 We leave the argument here, although clearly we have not yet proven all of Theorem 6.1. For instance, the third statement asserts strong rigidity, not just strong *reflection* rigidity. Moreover, we have yet to eliminate the word "reflection" from the second statement as well. Really all that is left to show is that if (W, S) is large-type and has no spikes as described in the theorem, then W is reflection independent, in which setting (strong) rigidity and (strong) reflection rigidity are equivalent. This is done in [Mühlherr and Weidmann (2002)], using a few more technical arguments involving chamber systems.

6.2   Circuits and centralizers

We now introduce a method of proof that will yield partial generalizations of some of the results in the previous section. The same method will be used again in Chapter 7 when investigating the structure of $\mathrm{Aut}(W)$ for certain groups W. As above, we will omit a number of the technical computations and proofs, focussing instead on the basic idea of the method and its applications.

We call the method "centralizer chasing". As the name suggests, we will make use of the form of the centralizer $C_W(W_T)$ for $T \subseteq S$ (in particular, for $T = \{s\}$). We use the circuits in the diagram $\mathcal{V}$ in much the same manner as before. We first compute an invariant (a group element) for each circuit in $\mathcal{V}$; we then splice the circuits together and use information about centralizers to show that the circuit invariants agree with one another, giving a well-defined element of the group W. This element will then give us the conjugating element needed to prove Theorem 6.3, for instance.

Let us first discuss the argument in its purest form.

6.2.1   *Walking around a circuit*

Suppose that (W, S) and (W, S') are systems for the same group W. Let $\mathcal{V}$ and $\mathcal{V}'$ denote the corresponding diagrams. For now, assume that (W, S) is two-dimensional (so that (W, S') is as well). From Proposition 4.1 it then follows that to each edge $[st]$ of $\mathcal{V}$ there is a unique edge $[s't']$ of $\mathcal{V}'$ such that $W_{\{s,t\}}$ and $W_{\{s',t'\}}$ are conjugate to one another. For an edge $[st]$, we define $W_{[st]} = W_{\{s,t\}}$.

Now consider a simple circuit $C = \{[s_1 s_2], [s_2 s_3], ..., [s_k s_1]\}$ in $\mathcal{V}$. That is, C is a collection of edges of $\mathcal{V}$, each of which shares a vertex with the previous edge, satisfying $s_i \neq s_j$ for $1 \leq i < j \leq k$. All arithmetic below will be done modulo k.

For each i, $1 \leq i \leq k$, Proposition 4.1 guarantees us an edge $[s''_{i-1} s'_i]$ and an element $w_i \in W$ such that $w_i W_{[s_{i-1} s_i]} w_i^{-1} = W_{[s''_{i-1} s'_i]}$. Considering possible generators for the dihedral group $W_{[s''_{i-1} s'_i]}$, we can modify w_i if needed so that

$$w_i s_{i-1} w_i^{-1} = s_{i-1}'' \text{ and } w_i s_i w_i^{-1} = z_{i-1,i} s_i' z_{i-1,i}^{-1}, \qquad (6.2)$$

for some $z_{i-1,i} = (s_i' s_{i-1}'')^j$, j satisfying $0 \leq j \leq [\frac{m-1}{4}]$ if $m = m_{s_{i-1}s_i}$ is even, and $0 \leq j \leq \frac{m-3}{2}$ if m is odd.

Of course, each vertex s_i lying on C appears in exactly two edges of C. From (6.2) applied to both of these edges, we see that s_i' and s_i'' are both conjugate to s_i, and therefore to each other. By Proposition 5.3, we can select a path $\{[s_i' t_1], ..., [t_r s_i'']\}$ in $\mathcal{V}'$ from s_i' to s_i'', every edge of which has an odd label.

Using the notation u_{st} and v_{st} from 2.2.2, we define

$$\bar{v}_i = v_{t_1 s_i'} v_{t_2 t_1} \cdots v_{s_i'' t_r}, \qquad (6.3)$$

so that $\bar{v}_i$ conjugates s_i'' to s_i'.

It now follows that

$$w_i s_i w_i^{-1} = z_{i-1,i} \bar{v}_i w_{i+1} s_i w_{i+1}^{-1} \bar{v}_i^{-1} z_{i-1,i}^{-1},$$

from which (after a bit of omitted calculation) we obtain $\bar{s}_i \in C_W(s_i')$ such that

$$w_{i+1} w_i^{-1} = \bar{v}_i^{-1} \bar{s}_i z_{i-1,i}^{-1}.$$

Denoting by x_i the word on the right-hand side of the last equation, we have

$$x_k x_{k-1} \cdots x_2 x_1 = w_1 w_k^{-1} w_k w_{k-1}^{-1} \cdots w_3 w_2^{-1} w_2 w_1^{-1} = 1. \qquad (6.4)$$

If we choose a geodesic representative for $\bar{s}_i$, each of the words $\bar{v}_i^{-1}$, $\bar{s}_i$, and $z_{i-1,i}^{-1}$ is as short as possible. Moreover, there cannot be a great deal of cancellation between these words when they are multiplied to obtain x_i. Of course, $z_{i-1,i}^{-1}$ may cancel with part of $(s_i' s_{i-1}'')^m$ if this last term appears at the end of $\bar{s}_i$. The only other cancellation occurs betweenm $\bar{v}_i^{-1}$ and $\bar{s}_i$. The first of these represents an "odd path" beginning at s_i'' and ending at s_i'; the second may begin either with another such path or with an "odd loop" based at s_i', as in Section 2.2.2. The first path may overlap with a long portion of the second. After cancellation of the corresponding common terms, we obtain

$$x_i = v_{\alpha_1 s_i''} v_{\alpha_2 \alpha_1} \cdots v_{s_i' \alpha_r} E_1 O_1 E_2 O_2 \cdots E_l O_l w(s_i', s_{i-1}'') \qquad (6.5)$$

where $w(s_i', s_{i-1}'') \in W_{[s_i' s'' i-1]}$, each E_i represents an "odd path" with an "even spike" in $C(s_i')$, and each O_i represents an "odd loop" in $C(s_i')$. (Here, $w(s_i', s_{i-1}'')$ is the subword appearing after possible cancellation with $z_{i-1,i}^{-1}$.)

Remark 6.5 The driving force of this method is the following: the form of $C(s_i')$ drastically limits the possible values of x_i, and therefore for the ratio between two conjugating elements w_i and w_{i+1}. Knowing that the product of these ratios must be simple enables us to go further; we will be able to show that each word x_i must be nearly trivial. This will give us, under the right circumstances, a single conjugating element, w_C, for a given circuit C. From there, it is merely a matter of piecing together the circuits of $\mathcal{V}$ in order to obtain rigidity results for the entire group W.

6.2.2 Centralizer chasing in even Coxeter systems

We first consider two different classes of even Coxeter systems, both of which consist entirely of two-dimensional systems. Having already obtained satisfactory results concerning reflection independence and rigidity in even groups, we focus on the issue of strong rigidity. In order to determine when an even group W is strongly rigid, we may as well assume that W is rigid and reflection independent (both necessary conditions for strong rigidity).

We have the following theorem, stated earlier as Theorem 3.12:

Theorem 6.3 *Let (W, S) be an even Coxeter system with connected diagram $\mathcal{V}$. If W is strongly rigid, then W is reflection independent and $\mathcal{V}$ contains no set of vertices J such that the following are both true:*

1. *The full subgraph Γ on the vertices $S \setminus J$ has at least 2 connected components, and*
2. *there are vertices s_1 and s_2 in different connected components of Γ and an element $w \in Z(W_J)$ such that $ws_1 \neq s_1 w$ and $ws_2 \neq s_2 w$.*

Moreover, if W is an even reflection independent Coxeter group whose diagram $\mathcal{V}$ has no such set J of vertices, then W is strongly rigid provided $\mathcal{V}$ has at least 3 vertices and satisfies one of the following conditions:

3. *$\mathcal{V}$ is of large-type, or*
4. *$\mathcal{V}$ contains no simple circuits of length 3 or 4.*

We call the sets J appearing above, *junctions*.

We will prove the assertion beginning "Moreover...". The reader may consult [Bahls (2004a)] for the remaining arguments.

Let (W, S) and (W, S') be even systems (with isomorphic even diagrams $\mathcal{V}$ and $\mathcal{V}'$) for the group W. Our work in applying the method of 6.2.1 is made considerably easier by the fact that there are no odd paths in $\mathcal{V}'$, and no two distinct generators are conjugate to one another. (In particular, $s_i' = s_i''$ for all i.) The form of x_i in (6.5) thus reduces to

$$x_i = u_{t_1 s_i'} u_{t_2 s_i'} \cdots u_{t_r s_i'} \epsilon_i z_{i-1,i}^{-1},$$

where $\epsilon_i \in \{1, s_i'\}$ and $z_{i-1,i}^{-1} = (s_{i-1}' s_i')^j$, $0 \le j \le [\frac{m-1}{4}]$, $m = m_{i-1,i}$. (Note that $j = 0$ when $m \in \{2, 4\}$.)

Because $s_i' = s_i''$ for all i, any simple circuit in $\mathcal{V}$ already corresponds to a simple circuit in $\mathcal{V}'$; we need only show that the conjugating words are the same for each edge in the circuit.

6.2.3 *The first case: $\mathcal{V}$ is a simple circuit*

We take the easiest case first, that in which $\mathcal{V}$ consists of a single simple circuit of length at least 5. Of course, we have already developed a method which demonstrates the strong rigidity of such systems (Exercise 7). We include the following discussion in order to motivate the method in more generality.

Our first step is the

Lemma 6.5 *We may assume $z_{i-i,i} = 1$ for all i.*

Proof. Apply to $x_k \cdots x_1 = 1$ the quotient map ν_i which identifies to 1 all vertices except s_i' and s_{i+1}'. Most of the terms of this product have trivial images, and what few remain must have as their product 1. From this one can show that $z_{i-1,i} \in \{1, s_i'\}$ (check!). If $z_{i-1,i} = s_i'$, this occurrence of s_i' can be absorbed by $\bar{s}_i$, and we can assume that $z_{i-1,i} = 1$. $\qquad\square$

Next we note that any letter s_i' can only appear in three of the terms in the product $x_k \cdots x_1$; namely, in x_{i-1}, x_i, and x_{i+1}. (This follows from the fact that $\mathcal{V}'$ is a simple circuit.) Thus we really only need to consider cancellation between these three words.

Considering first the product $x_{i+1} x_i$, we may cancel up to $2m_{i,i+1}$ letters (this maximum occurs when x_{i+1} ends with $u_{i,i+1} s_{i+1}'$ and x_i begins with $u_{i+1,i} s_i'$. Similarly, at most $2m_{i-1,i}$ letters cancel in the product $x_i x_{i-1}$. Finally, there can be cancellation between x_{i+1} and x_{i-1}, but only if $x_i = 1$,

x_{i+1} ends with $u_{i,i+1}$, and x_{i-1} begins with $u_{i,i-1}$. (In this case a single pair of letters cancels.)

We now use van Kampen diagrams to show that each x_i is "short":

Lemma 6.6   *Each generator $u_{i+1,i}$ and $u_{i-1,i}$ of $C(s_i')$ can occur at most once in x_i.*

Proof.   We let Δ be a van Kampen diagram with boundary label $\partial\Delta$ equal to the word resulting from $x_k \cdots x_1$ after all free reduction has been performed. (See Section 1.4 for details regarding van Kampen diagrams.) Assume that $u_{i-1,i}$ appears more than once, so that $u_{i-1,i}u_{i+1,i}u_{i-1,i}$ is a subword of x_i. (The other possibility is entirely symmetric.) After freely reducing the product $x_k \cdots x_1$, at least $u_{i-1,i}u_{i+1,i}$ remains.

We construct bands in Δ, much as was done in the proof of the Deletion Condition in Section 1.4. Consider two bands, one beginning at any edge labeled s_{i-1}' lying in the word $u_{i-1,i}$ from the previous paragraph, the other at any edge labeled s_{i+1}' in the word $u_{i+1,i}$. The other ends of these bands must be edges labeled s_{i-1}' and s_{i+1}', respectively. It is not difficult to see that in order for these bands to exist, they must cross each other, so that the order of $s_{i-1}'s_{i+1}'$ must be finite, which is not the case. This gives a contradiction. $\square$

Thus there is a uniform bound on the length of the words x_i (so in particular there are only finitely many choices for each x_i). Now, because the product of all of the x_i's is trivial and because each x_i admits cancellation only with nearby x_j's, there must be a good deal of "telescoping" in the product $x_k \cdots x_1$. This is made precise by the following proposition:

Proposition 6.9   *For every i, $2 \leq i \leq k-2$, the element of W represented by $x_k \cdots x_i$ is equal in W to a word of the form xv_i, where*

1. x is a word in the letters s_1' and s_k', and $x \in C(s_1') \cap C(s_k')$, and
2. v_i is a word in the letters s_i' and s_{i-1}', and $v_i \in C(s_i') \cap C(s_{i-1}')$.

Proof.   This can be proven using induction on i, and the fact that (due to Lemma 6.6) there are only finitely many choices for each x_i that need to be considered. The details are left as an exercise. $\square$

It is now left to the reader to show that the word $w = w_1^{-1}x$ (for x as in Proposition 6.9) has the property that $ws_i'w^{-1} = s_i$ for every $i = 1, ..., k$. This shows that W is strongly rigid.

6.2.4 *The general case*

Now let us turn to the case in which $\mathcal{V}$ consists of more than a simple circuit. The first step is to prove the

Lemma 6.7 *Let* (W, S) *be a reflection independent even Coxeter system whose diagram* $\mathcal{V}$ *does not contain a junction, as in Theorem 6.3. Then either* W *is dihedral or* $\mathcal{V}$ *contains no vertices of valence* 1.

An immediate consequence of this fact is the following:

Lemma 6.8 *Let* (W, S) *be a reflection independent even Coxeter system whose diagram* $\mathcal{V}$ *contains neither a junction nor a simple circuit of length less than* 5. *Then if* W *is not dihedral, every vertex of* $\mathcal{V}$ *lies on some simple circuit* $C = \{[s_1 s_2], ..., [s_k s_1]\}$ *of* $\mathcal{V}$ *which satisfies the following condition:*

(*) *for any vertices* s_i *and* s_j *on* C *which are no adjacent on* C, $\mathcal{V}$ *does not contain a path of length at most* 2 *from* s_i *to* s_j.

Any simple circuits satisfying the condition (*) will be called a *regular* circuit. We prove Lemma 6.7 and leave the simple proof of Lemma 6.8 as an exercise.

Proof. Let $[st]$ be an edge for which the vertex t has valence 1. Using the results of Chapter 5 it is easy to show that $[st]$ cannot have label $2(2k+1)$ for any $k \geq 0$. (Were this to occur, (W, S) would not be reflection independent; *cf.* Remark 6.4.)

Therefore $[st]$ has label $4k$, $k \geq 1$. We may assume there is some edge other than $[st]$ (otherwise W is dihedral). In this case, s separates $\mathcal{V}$ into more than one component. If there were a vertex $t' \in S$ such that $st' \neq t's$, then $\{s\}$ would be a junction, which we have assumed is not the case. Thus $t' \in S \setminus \{s, t\} \Rightarrow st' = t's$. But then for any $t' \in S \setminus \{s, t\}$, $\mathrm{St}(t') \subseteq \mathrm{St}_2(s)$, contradicting (W, S)'s reflection independence, by Proposition 5.9. $\qquad\Box$

Suppose now that $C = \{[s_1 s_2], ..., [s_k s_1]\}$ is a regular circuit in $\mathcal{V}$, and that $\mathcal{V}$ has neither junctions nor simple circuits of length less than 5 (so $k \geq 5$, in particular). We may obtain a circuit $C' = \{[s_1' s_2'], ..., [s_k' s_1']\}$ in $\mathcal{V}'$ and words $x_i = w_{i+1} w_i^{-1}$ and $z_{i-1,i}$ must as above. The only real difficulty now lies in the fact that x_i could contain letters which do not lie on the circuit C'.

Lemma 6.9 *If* $t' \in S'$ *does not lie on* C', *then* t' *cannot appear in any word* x_i.

Proof. Suppose that t' appears in x_i. By regularity of C', t' cannot be adjacent to any letter s'_j, $|j - i| > 2$ (here subtraction is modulo k). However, since $\mathcal{V}'$ contains no circuits of length less than 5, t' can only be adjacent to s'_i, so after free cancellation in the product $x_k \cdots x_1$, this letter t' remains in the word x_i. Since $x_k \cdots x_1 = 1$, each occurrence of t' must cancel. Since all such cancellations occur between letters of x_i, this contradicts the geodesity of x_i. Thus no such t' can appear. $\qquad\square$

Thus there is no real difference between the present case and the case in which $\mathcal{V}$ is a circuit. Therefore given a regular circuit C in $\mathcal{V}$, there exists a regular circuit C' in $\mathcal{V}'$ and a single word $w \in W$ such that w conjugates each edge of C to the corresponding edge of C'.

Given two regular circuits C_1 and C_2 in $\mathcal{V}$, let Γ denote the set of vertices which lie on both C_1 and C_2. If w_i is the conjugating word for C_i, then for any vertex s in Γ, $w_1^{-1}w_2 \in C(s)$. Therefore if Γ is large, the conjugating words w_1 and w_2 are very similar, since intersection of centralizers of vertices are quite small (see Theorem 2.6).

For instance, suppose that C_1 and C_2 share more than three vertices. It is not hard to see that $w_1 = w_2$ must hold in this case. This is true also if $|\Gamma| = 3$ and the three vertices in Γ are not consecutive on C_i, or if they are consecutive but at least one of the edges has label greater than 2. There are three cases to consider then: $|\Gamma| = 1$, $|\Gamma| = 2$, and $|\Gamma| = 3$ and both edges are labeled 2. (The second case divides naturally into 2 subcases.)

We sketch the proof in the case in which Γ is an edge $[st]$ with label $m > 2$. The other cases are proven in similar fashion.

The only undesirable possibility here is $w_1^{-1}w_2 = (st)^{\frac{m}{2}}$. In order to show that this cannot be, we will construct a "chain" of adjacent regular circuits which connects C_1 to C_2 without passing through $[st]$. (*Cf.* the proof of Proposition 6.7.)

Why can this be done? Since $\mathcal{V}$ contains no junctions, removing s and t (and all incident edges) does not disconnect the diagram $\mathcal{V}$. Therefore we can find some path, p, which contains neither s nor t, and which connects a vertex in $C_1 \setminus \{s, t\}$ to a vertex in $C_2 \setminus \{s, t\}$. Concatenating p with the appropriate subpath q of $C_1 \cup C_2$, we obtain a new circuit, D. (See Figure 6.5. Note that, as in this figure, p may contain vertices from $C_1 \cup C_2 \setminus (\{s, t\})$.)

The new circuit D may not be regular. (If it is, we are done: if $w \in W$ is the conjugating word for D, then $w^{-1}w_1 \cdot w_1^{-1}w_2 \cdot w_2^{-1}w = 1$, and it is not hard to show that neither the first nor the last term in this product

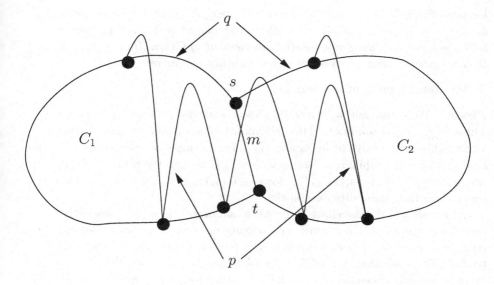

Fig. 6.5   Overlapping regular circuits

can cancel the word $(st)^{\frac{m}{2}}$ appearing in $w_1^{-1}w_2$.) However, by choosing the points at which p intersects C_1 and C_2 carefully, we can construct D which may fail to be regular only because one of the letters s and t can be connected to some other vertex p_i on p by a path of length at most 2.

By adding such short paths, we subdivide D into (necessarily shorter) circuits, which themselves can be further subdivided in a similar fashion, provided they are not regular. Eventually every circuit created by this process of subdivision is regular. From this collection we may obtain a chain of regular circuits $C_1, D_1, ..., D_k, C_2$, where any two consecutive circuits have small intersection (so that the ratio of consecutive conjugating words is small). Arguing as in the last paragraph's parenthetical remark, we can conclude that $w_1^{-1}w_2 \neq (st)^{\frac{m}{2}}$ must hold. The reader is urged to consult [Bahls (2004a)] for details.

The case in which we allow circuits of length 3 and 4, but not edges with label 2, is proven similarly. Instead of working with regular circuits, as defined above, we work with chord-free circuits.

As in the previous case, the principal difficulty lies in the fact that a word x_i may contain letters which do not lie on the given circuit. We claim, however, that this cannot be:

Lemma 6.10 *Suppose that $\mathcal{V}$ contains neither junctions nor edges labeled 2. Let $C = \{[s_1 s_2], ..., [s_k s_1]\}$ be a chord-free circuit in $\mathcal{V}$, and for each i let v_i be a geodesic word representing an element of $C(s_i)$. If $v_1 \cdots v_k = 1$, then no word v_i contains a letter which does not appear on C.*

We sketch a proof of this lemma.

Proof. Were this not so, we could choose a trivial product $v_1 \cdots v_k$ for which $\sum_{i=1}^{k} |v_i|$ is minimal. This minimality assumption outlaws certain forms of the consecutive terms v_i and v_{i+1}. For example, we cannot have v_i ending with $u_{i+1,i}$ while v_{i+1} begins with $u_{i,i+1}$, as then the product $v_i v_{i+1}$ would have the form $v_i' s_i \cdot s_{i+1} v_{i+1}'$, for $v_i' \in C(s_i)$ and $v_{i+1}' \in C(s_{i+1})$. This contradicts the minimality assumption.

After allowable cancellation between small portions of consecutive terms, we readily obtain a word representing $u_1 \cdots u_k = 1$ which does not contain as a subword more than half of a relator word $(st)^{m_{st}}$. This contradicts Tits's solution to the Word Problem (see Section 2.4). (Those who know something about small cancellation theory might also notice that W is a $C'(\frac{1}{6})$ group; this fact yields a slightly different contradiction.) $\qquad\square$

Applying Lemma 6.10 to the words x_i, we see that no x_i can contain letters which do not lie on C.

There is one more minor obstacle to take care of. Above, we only proved strong rigidity of simple circuits of length at least 5, while now we must consider shorter circuits. Exercise 10 asks the reader to fill in the remaining details.

We can now apply the method of "overlapping circuits" outlined above in order to find a single conjugating word for all of $\mathcal{V}$, just as before.

Remark 6.6 In the more general setting of arbitrary two-dimensional Coxeter systems, the method of centralizer chasing, suitably modified, should still work. In this case, the words x_i may have more complicated forms, but arguments similar to the above should obtain.

6.3 Exercises

1. Let (W, S) be a Coxeter system, and recall the definitions of 6.1.1. Prove that $G \leq W$ is a finite subgroup of W if and only if G leaves invariant some spherical residue in the chambers of the chamber system $Ch(W, S)$.

2. With the definitions of 6.1.1, prove that a reflection $t \in R(S)$ leaves

invariant the residue $R_T(u)$ if and only if both roots of t intersect $R_T(u)$ nontrivially.

3. Prove that any root (as defined in 6.1.1) corresponding to a reflection $t \in R(S)$ is convex.

4. Let r_1, r_2, r_3 be distinct reflections in $R(S)$ such that $r_1 r_2$ and $r_1 r_3$ both have infinite order. Let $H(r_1, r_2) = -H(r_1, r_3)$. Prove that $r_2 r_3$ has infinite order.

5. Throughout this exercise let $\{H, H'\}$ be a 2-element geometric set of roots, as defined in 6.1.2.

    a. Suppose that $\{-H, H'\}$ is another 2-element geometric set of roots. Prove the $r_H r_{H'} = r_{H'} r_H$, and that any of the 4 sets $\{\pm H, \pm H'\}$ is a geometric set of roots.

    b.   Suppose that additionally $r_H H \cap r_{H'} H' \neq \emptyset$. Prove that $W(\{r_H, r_{H'}\})$ is a finite subgroup of W, and that $\{r_H H, r_{H'} H'\}$ is a geometric set of roots.

    c. Suppose that $r_H r_{H'}$ has infinite order. Prove that $H = H(r_H, r_{H'})$, $H' = H(r_{H'}, r_H)$ and that $H \cap H'$ is a fundamental domain for the action of the group $W(\{r_H, r_{H'}\})$ on the chambers of $Ch(W, S)$. Prove that there is no *other* 2-element geometric set of roots associated to $\{r_H, r_{H'}\}$.

6. Prove that w_i can be chosen as asserted in (6.2) and the paragraph that follows.

7. Use Theorem 4.4 to prove that any two-dimensional Coxeter system whose diagram $\mathcal{V}$ consists of a simple circuit is strongly rigid.

8. Complete the proof of Proposition 6.9 by filling in the details of the indicated induction.

9. Prove Lemma 6.8.

10. Give an example to show why the method of proof adopted in 6.2.2 cannot be used to show that all simple circuits of length 4 are strongly rigid. Show that under the additional assumption that no edges are labeled 2, the proof goes through much as before.

Chapter 7

Refinements and generalizations: automorphisms and Artin groups

We have now detailed a number of results concerning the uniqueness of presentation of Coxeter groups. In this chapter, we consider ways in which these results can be generalized.

We first take a closer look at the group $\mathrm{Aut}(W)$ of automorphisms of W. (Recall that both rigidity and strong rigidity are statements about the robustness of the groups $\mathrm{Aut}(W)$ and $\mathrm{Inn}(W)$.) A greater understanding of the structure of Aut(W) will certainly lead to a greater understanding of W itself.

After analyzing $\mathrm{Aut}(W)$ for a number of Coxeter groups, we will examine briefly the extent to which we can apply the results from this and the previous chapters to a class of groups other than Coxeter groups, the Artin groups.

7.1 Automorphisms of right-angled groups

As before, we begin our investigation by considering the simplest of Coxeter groups. Our first goal is to understand $\mathrm{Aut}(W)$ when W is right-angled. This will be done by describing $\mathrm{Aut}(W)$ in terms of certain of its subgroups, and then by explaining what those subgroups look like.

7.1.1 *The group* $\mathrm{Spe}(W)$

In most Coxeter groups (including most right-angled groups), the center $Z(W)$ of W is trivial. Therefore, in most instances $W \cong \mathrm{Inn}(W)$, so the group of inner automorphisms is very easy to understand. This group is closely related to another subgroup, $\mathrm{Spe}(W)$, of $\mathrm{Aut}(W)$ with which we deal extensively below.

Given *any* group G, $\mathrm{Aut}(G)$ acts on the collection of conjugacy classes of involutions in G, in the obvious fashion:

$$\alpha([g]) = [\alpha(g)]$$

where $[g]$ is the conjugacy class of the involution $g \in G$.

It is clear that the subgroup

$$\mathrm{Spe}(G) = \{\alpha \in \mathrm{Aut}(G) \mid \alpha([g]) = [g] \text{ for all involutions } g \in G\} \qquad (7.1)$$

of *special automorphisms* of G satisfies $\mathrm{Inn}(G) \leq \mathrm{Spe}(G) \leq \mathrm{Aut}(G)$.

In the case of Coxeter groups, we note that R.W. Richardson (in [Richardson (1982)]) has proven that any Coxeter group W has only finitely many conjugacy classes of involutions, implying that $\mathrm{Spe}(W)$ has finite index in $\mathrm{Aut}(W)$. (This is also true of a number of other groups, notably of word hyperbolic groups.) Therefore, an understanding of $\mathrm{Spe}(W)$ will permit an understanding of $\mathrm{Aut}(W)$. We can make the relationship between these groups more precise.

As in 1.3.3, denote by $\mathcal{S} = \mathcal{S}(W)$ the collection of spherical subsets of $\mathcal{V}$ (which in the right-angled case coincides with the set of simplices of $\mathcal{V}$). We may define a partial product, $\cdot$, on $\mathcal{S}$ by $T_1 \cdot T_2 = (T_1 \setminus T_2) \cup (T_2 \setminus T_1)$ for $T_i \in \mathcal{S}(S)$ (so $T_1 \cdot T_2$ is the symmetric difference of the two sets). It is easy to show that $(\mathcal{S}, \cdot \,)$ is a finite groupoid. Thus its automorphism group $\mathrm{Aut}(\mathcal{S})$ is also finite. Each $\beta \in \mathrm{Aut}(\mathcal{S})$ induces an automorphism $\hat{\beta}$ on W itself by

$$\hat{\beta}(s) = \prod_{t \in \beta(\{s\})} t. \qquad (7.2)$$

The reader should check that this does define an automorphism.

Remark 7.1   There is no fully standard use of the term *groupoid* in the mathematical literature. For our purposes, a groupoid is a set G with a partial product, $\cdot$, and a function $\mathrm{inv} : G \to G$, such that

1. if $g_1 \cdot g_2$ and $g_2 \cdot g_3$ are both defined, then both $(g_1 \cdot g_2) \cdot g_3$ and $g_1 \cdot (g_2 \cdot g_3)$ are defined and are equal,
2. for every $g \in G$, $g \cdot \mathrm{inv}(g)$ and $\mathrm{inv}(g) \cdot g$ are both defined, and
3. whenever $g_1 \cdot g_2$ is defined, $g_1 \cdot g_2 \cdot \mathrm{inv}(g_2) = g_1$ and $\mathrm{inv}(g_1) \cdot g_1 \cdot g_2 = g_2$.

With $\cdot$ defined on $\mathcal{S}$ as above, $\mathrm{inv}(T) = T$ for all $T \in \mathcal{S}$, and the empty set $\emptyset$ serves as a groupoid identity: $\emptyset \cdot T = T \cdot \emptyset = T$ for all $T \in \mathcal{S}$.

Our first goal is to prove the following theorem:

Theorem 7.1 *Let W be a right angled Coxeter group with diagram $\mathcal{V}$. Then $\mathrm{Aut}(W)$ is a semidirect product of $\mathrm{Aut}(S)$ by $\mathrm{Spe}(W)$. Moreover, if $\mathcal{V}$ contains no triple of distinct, pairwise non-adjacent vertices, then $\mathrm{Spe}(W) = \mathrm{Inn}(W)$.*

Theorem 7.2 below will provide a refinement by telling us exactly what $\mathrm{Spe}(W)$ looks like.

We now sketch a proof of the above result. The first statement is not difficult to verify; the second requires a bit of work.

First note that $\mathrm{Spe}(W)$ consists precisely of those automorphisms α which induce the identity map on the abelianization W/W' of W. (Prove!)

Suppose $|S| = n$. The group W/W' is isomorphic to $\mathbb{Z}_2^n$. Denote by $\bar{S}$ the subgroup of W/W' corresponding to the groupoid S. (Thus $\bar{S}$ consists of the elements $(\epsilon_1, ..., \epsilon_n)$, $\epsilon_i \in \{0, 1\}$, for which $\{s_i \mid \epsilon_i = 1\} \in S$).

Any automorphism $\alpha \in \mathrm{Aut}(W)$ induces a map $\bar{\alpha} \in \mathrm{Aut}(W/W')$ in the natural fashion, and $\bar{\alpha}$ leaves stable the set $\bar{S}$ (by Lemma 1.2). Thus $\bar{\alpha}$ induces an automorphism of $\bar{S}$, which we also denote by $\bar{\alpha}$. The kernel of the map $\alpha \mapsto \bar{\alpha}$ is precisely the group $\mathrm{Spe}(W)$.

Furthermore, the map which associates with a given automorphism β of $\bar{S}$ (and therefore of S itself) the map $\hat{\beta}$ described in (7.2) above is a section to the map $\alpha \mapsto \bar{\alpha}$. Therefore there is a split exact sequence witnessing the desired semidirect product.

We now prove the second statement ("Moreover...") of Theorem 7.1 by arguing the contrapositive. (Since $\mathrm{Inn}(W) \leq \mathrm{Spe}(W)$, $\mathrm{Spe}(W) \neq \mathrm{Inn}(W)$ implies that $\mathrm{Spe}(W) \backslash \mathrm{Inn}(W) \neq \emptyset$.) Our first lemma handles a special case:

Lemma 7.1 *Suppose that the diagram $\mathcal{V}$ has exactly two maximal simplices, σ_1 and σ_2. Then $\mathrm{Spe}(W) = \mathrm{Inn}(W)$.*

Proof. [Tits (1988)] contains a proof of this lemma which uses the action of W upon its Coxeter complex. A different argument applies the method used below to prove Theorem 3.16. The reader is encouraged both to supply the details of this latter argument, and to modify this argument in order to prove Lemma 7.1. □

Let's consider another special case of Theorem 7.1. Given $s \in S$, define $s^{\perp}$ by $s^{\perp} = \{t \in S \mid m_{st} = \infty\}$. (In terms of $\mathcal{V}$, $s^{\perp}$ is the collection of vertices which are not adjacent to s; it is a sort of "co-link" of s.)

Lemma 7.2 *Suppose that for some vertex $s \in \mathcal{V}$, $s^{\perp}$ can be expressed as* *the disjoint union of two nonempty sets S_1 and S_2 such that no element of S_1 is adjacent to any element of S_2. Then $\mathrm{Spe}(W) \neq \mathrm{Inn}(W)$.*

Proof.   It is left to the reader to check that all of the defining relations of the group W remain satisfied when each element t of S_1 is replaced by sts. Thus the map α defined by $\alpha(t) = t$ for $t \notin S_1$ and $\alpha(t) = sts$ for $t \in S_1$ is an automorphism (since it is an involution). Suppose that $\alpha \in \mathrm{Inn}(W)$, so $\alpha(t) = wtw^{-1}$ for some w, for all $t \in S$. Then $sw \in C(W_{S_1})$, so w is in the group generated by $C(W_{S_1})$ and s itself.

However, $w \in C(W_{S \setminus S_1})$ as well. Because $s \notin C(W_{S \setminus S_1})$ (this from $S_2 \neq \emptyset$), it follows that w centralizes all of W. However, $Z(W)$ is clearly trivial, so this is a contradiction. $\square$

Now, yet another special case. The following lemma will allow us to show inductively that the configuration forbidden by Theorem 7.1 must appear if we assume $\mathrm{Spe}(W) \neq \mathrm{Inn}(W)$.

Lemma 7.3   *Suppose that the vertex s in $\mathcal{V}$ does not appear in any triple $\{s_1, s_2, s_3\}$ of distinct, pairwise non-adjacent vertices. Suppose also that $\mathrm{Spe}(W) \neq \mathrm{Inn}(W)$, but that $\mathrm{Spe}(C(s)) = \mathrm{Inn}(C(s))$. Then $\mathcal{V}$ must contain a full subdiagram of the form shown in Figure 7.1, where if x lies in $\mathrm{Lk}(r) \cap \mathrm{Lk}(s)$, then $x \in \mathrm{Lk}(t)$.*

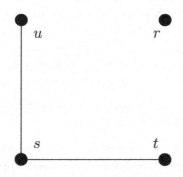

Fig. 7.1   The necessary configuration in Lemma 7.3

Once Lemma 7.3 in shown, the proof of Theorem 7.1 is easy: we induct on $|S|$. Suppose $\mathcal{V}$ contains no triple of distinct pairwise non-adjacent vertices, and let $s \in S$. By the inductive hypothesis, $\mathrm{Spe}(C(s)) = \mathrm{Inn}(C(s))$. Lemma 7.3 now implies that $\mathcal{V}$ must contain the forbidden configuration unless $\mathrm{Spe}(W) = \mathrm{Inn}(W)$, completing the proof. So, we prove Lemma 7.3:

Proof. Since $\mathrm{Spe}(W) \neq \mathrm{Inn}(W)$, we can choose $\alpha \in \mathrm{Spe}(W) \setminus \mathrm{Inn}(W)$. Composing α with an inner automorphism, if necessary, we may suppose that $\alpha(s) = s$. Furthermore, since α maps the group $C(s)$ to itself, and since $\mathrm{Spe}(C(s)) = \mathrm{Inn}(C(s))$ is assumed true, we may compose α with an inner automorphism of $C(s)$ and assume that α fixes (pointwise) the star of s, $\mathrm{St}(s)$.

Pick and fix such an automorphism α for which the number of vertices satisfying $\alpha(t) = t$ is maximal among such maps. Since $\alpha \neq \mathrm{id}_W$, there is some vertex r such that $\alpha(r) \neq r$. Our hypotheses imply that rs has infinite order, so $[rs]$ is not an edge in $\mathcal{V}$.

Now let $S' \subseteq S$ denote the collection of vertices x of $\mathcal{V}$ which are adjacent to every vertex y which is in turn adjacent to both r and s. Let $\mathcal{V}'$ be the full subdiagram of $\mathcal{V}$ induced by the vertices S'. (That is, S' generates the centralizer of the centralizer of $W_{\{r,s\}}$. In particular, $W_{\{r,s\}} \leq W_{S'}$.) Since α fixes $C(s)$, it also fixes $C(W_{\{r,s\}}) \leq C(s)$, and it therefore maps the group $W_{S'}$ onto itself. If $r \in \{s_1, s_2, s_3\} \subseteq S'$ where the s_i are distinct and pairwise non-adjacent, this triple gives the desired configuration. (Please check this!)

So we suppose that r does not appear in such a triple. In this case it is not hard to see that $\mathcal{V}'$ has the form described in the statement of Lemma 7.1. (Namely, $\mathcal{V}'$ has the two maximal simplices $\sigma_1 = \mathrm{St}(r) \cap \mathcal{V}'$ and $\sigma_2 = \mathrm{St}(s) \cap \mathcal{V}'$.) Thus, for some $w \in W_{S'}$, $wuw^{-1} = \alpha(u)$ for all $u \in W_{S'}$, and it follows that w can be written as a product of generators in S' that commute with s; that is, $w \in W_{\sigma_2}$. We lose no generality in assuming that each such generator appearing in w does not commute with r. (Any letter of $\sigma_1 \cap \sigma_2$ appearing in w can be removed from w, so that $w \in W_{\sigma_2 \setminus \sigma_1}$.)

Now there is some letter $u \in S$ such that $\alpha(u) = u$ and $wuw^{-1} \neq u$; otherwise, composing α on the left with conjugation by w^{-1} would yield a non-inner automorphism which contradicts the maximality assumption on α. Indeed, in this case $w^{-1}\alpha(u)w = w^{-1}uw = u$ for all u fixed by α, and moreover $w^{-1}\alpha(r)w = w^{-1}wrw^{-1}w = r$.

Suppose that $ur = ru$. Then $w^{-1}\alpha(u)w$ and $w^{-1}\alpha(r)w = r$ commute. Since $w^{-1}uw$ lies in $W_{(\sigma_2 \setminus \sigma_1) \cup \{u\}}$ and none of the letters of $\sigma_2 \setminus \sigma_1$ commute with r, $w^{-1}uw = u$ after all. Thus $ur \neq ru$.

At last, choose t to be any letter of $\sigma_2 \setminus \sigma_1$ different from u (which must exist, since $wu \neq uw$). It is left to the reader to show that these vertices provide the configuration pictured in Figure 7.1. $\square$

7.1.2   *What does* Spe(*W*) *look like?*

We now describe the structure of Spe(W) completely (Theorem 7.2) in case W is right-angled. The proof of the theorem below (which can be found in [Mühlherr (1998)]) is quite long and technical, and we will omit it.

Let (W, S) be a right-angled system, and define $s^\perp$ as before, for $s \in S$. We call a subset $T \subseteq s^\perp$, *s-admissible* if it is either empty or a connected component of the full subgraph of $\mathcal{V}$ induced by the set $s^\perp$. A pair (s, T) is called *admissible* if T is s-admissible. (For example, the s-admissible sets are indicated in Figure 7.2.)

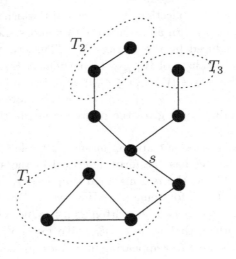

Fig. 7.2 s-admissible sets T_i

Given an admissible pair (s, T), define the automorphism (involution, in fact) σ_{sT} by $\sigma_{sT}(t) = sts$ if $t \in T$ and $\sigma_{sT}(t) = t$ otherwise. Clearly each such automorphism lies in Spe(W), and furthermore there are enough of these automorphisms to generate Spe(W).

What relations are there between such maps?

If $T = \emptyset$, then

$$\sigma_{sT} = \mathrm{id}_W. \tag{7.3}$$

Also, if we first act by σ_{sT_1} and then by σ_{sT_2}, we undo the action of the first map on all those elements which lie in $T_1 \cap T_2$:

$$\sigma_{sT_2}\sigma_{sT_1} = \sigma_{sT}, \tag{7.4}$$

where $T = (T_1 \setminus T_2) \cup (T_2 \setminus T_1)$.

Some of these automorphisms commute with one another:

$$\sigma_{s_1T_1}\sigma_{s_2T_2} = \sigma_{s_2T_2}\sigma_{s_1T_1} \tag{7.5}$$

if $s_1s_2 = s_2s_1$ or if $T_1 \cup \{s_1\}$ and $T_2 \cup \{s_2\}$ are disjoint.

Finally, we register how the "maximal" automorphism $\sigma_{ss^\perp}$ behaves:

$$\sigma_{tT}\sigma_{ss^\perp} = \sigma_{tt^\perp}\sigma_{ss^\perp}\sigma_{tt^\perp}\sigma_{tT} \tag{7.6}$$

if $s \in T$, and

$$\sigma_{tT}\sigma_{ss^\perp} = \sigma_{ss^\perp}\sigma_{tT} \tag{7.7}$$

otherwise. We claim now that all relations in $\mathrm{Spe}(W)$ are a consequence of these. Namely, we have

Theorem 7.2 *Given a right angled Coxeter system* (W, S), *the group* $\mathrm{Spe}(W)$ *of special automorphisms has a presentation with generators* σ_{sT}, *where* (s, T) *ranges over all admissible pairs, and relations defined as above. Moreover,* $\mathrm{Spe}(W) = \mathrm{Inn}(W)$ *if and only if for all* $s \in S$, $s^\perp$ *is connected.*

7.2 Coxeter systems with $\mathcal{V}$ complete

We return now to very geometric arguments.

In this section we assume that for all $s, t \in S$, $m_{st} < \infty$. That is, $\mathcal{V}$ is complete. (In [Howlett, *et al.* (1997)], such a group is said to have *no infinite bonds*.) These groups are freely indecomposable, and are in fact one-ended (as shown by the methods of M. Mihalik and S. Tschantz highlighted in Chapter 2). Therefore these groups are in a certain sense "small".

In this section we will prove the following:

Theorem 7.3 *Let* (W, S) *be a Coxeter system for which* $\mathcal{V}$ *is a complete graph. Then* $|\mathrm{Out}(W)|$ *is finite.*

The proof will make use of the characterization of W as a group generated by reflections in some vector space V, given in 1.2.2. We need to examine carefully the Tits cone defined in 1.5.2. We will also require the notion of *dominance* (developed by Brink and Howlett in [Brink and Howlett (1993)]) in the root system corresponding to V. More will be said about this below (see Exercise 1.21 for the definition of the dominance order).

7.2.1 *A chain of subgroups of* Aut(W)

We prove Theorem 7.3 by showing that $\text{Inn}(W)$ has finite index inside a certain subgroup $\text{Aut}_V(W)$ of $\text{Aut}(W)$, and that $\text{Aut}_V(W)$ has finite index in $\text{Aut}(W)$ itself. Let $S = \{s_i\}_{i \in I}$ for $|I| < \infty$. Also, let V be a vector space with basis $\{\alpha_i \mid i \in I\}$ and a bilinear form $\langle \cdot, \cdot \rangle$ defined as in 1.2.2. (That is, s_i acts on $v \in V$ by $s_i(v) = v - 2\langle \alpha_i, v \rangle \alpha_i$. In this section we use angle brackets instead of parentheses to avoid confusion with other notation which will appear below.) The set $\mathcal{S}(S)$ of spherical subsets of S is finite, since $|S| < \infty$.

We next define $\text{Inv} = \text{Inv}(W)$ to be the set of involutions (elements of order 2) in W, and $P = P(W, S)$ to be the set of pairs

$$\{(x, y) \mid x, y \in \text{Inv and } xy \text{ has finite order}\}.$$

The group W itself acts on P by conjugation:

$$w \cdot (x, y) = (wxw^{-1}, wyw^{-1}).$$

$\text{Aut}(W)$ acts on P as well, in such a way that each $\phi \in \text{Aut}(W)$ permutes the W-orbits of P. From Lemma 1.2 it follows that there are only finitely many W-orbits in P (prove this!).

Define $\text{Ker} = \text{Ker}(W)$ to be the kernel of the action of $\text{Aut}(W)$ on the W-orbits of P. (That is, $\phi \in \text{Ker}$ if and only if for every pair $x, y \in \text{Inv}$, $(\phi(x), \phi(y)) = (wxw^{-1}, wyw^{-1})$ for some $w \in W$.) Since the number of W-orbits in P is finite, it follows that $[\text{Aut}(W) : \text{Ker}]$ is finite.

Remark 7.2 $\text{Ker}(W)$ and $\text{Spe}(W)$, though defined in roughly similar fashion, are not necessarily the same. The reader is encouraged to show this by constructing an example.

Now we define the group $\text{Aut}_V(W)$ mentioned above. As the notation suggests, this group consists of the automorphisms of W induced by orthogonal transformations on the vector space V. That is, given an orthogonal transformation $\tau : V \to V$, define $\tau(s_i) = r_{\tau(\alpha_i)}$, where (as in Chapter 1) for any vector $v \in V$, r_v represents the element of W taking v to $-v$ and fixing the orthogonal complement $v^{\perp}$. (Explicitly writing the generators as reflections $r_i = r_{\alpha_i}$, we may also write $\tau(r_i) = \tau r_i \tau^{-1}$.)

We will show first that $\mathrm{Aut}_V(W)$ has finite index in $\mathrm{Aut}(W)$, and then that $\mathrm{Inn}(W)$ has finite index in $\mathrm{Aut}_V(W)$, from which Theorem 7.3 follows.

7.2.2 *The first step*

Proposition 7.1 $[\mathrm{Aut}(W) : \mathrm{Aut}_V(W)]$ *is finite.*

Proof. Let (W, S) have no infinite bonds, and let $S = \{s_i\}_{i \in I}$. Define Inv, P, and Ker as before.

From the definition of Ker and the fact that (W, S) has no infinite bonds, it follows that to each $\phi \in \mathrm{Ker}$ and each pair $s, t \in S$, there exists $w = w(\phi, s, t) \in W$ satisfying

$$\phi(s) = wsw^{-1} \text{ and } \phi(t) = wtw^{-1}.$$

Moreover, these elements can be chosen so that $w(\phi, s, t) = w(\phi, t, s)$ holds.

Fix $\phi \in \mathrm{Ker}$ for the moment. For $s, t \in S$, let $w = w(\phi, s, s)$ and $w' = w(\phi, s, t)$. Then

$$r_{w\alpha_s} = wsw^{-1} = \phi(s) = w'sw'^{-1} = r_{w'\alpha_s},$$

where r_v is defined as before for each vector $v \in V$. The upshot of this equation is that $w'\alpha_s = \pm w\alpha_s$. From this (and the fact that the action of each $w \in W$ on V is orthogonal) it follows that

$$\langle w(\phi, s, s)s, w(\phi, t, t)t \rangle = \epsilon(s, t)\langle s, t \rangle$$

for some $\epsilon = \epsilon(\phi, s, t) \in \{\pm 1\}$ depending on ϕ, s, and t.

We need the following lemma.

Lemma 7.4 *If* $\phi, \phi' \in \mathrm{Ker}$ *satisfy* $\epsilon(\phi, s, t) = \epsilon(\phi', s, t)$ *for all* $s, t \in S$, *then* $\phi' = \tau\phi$ *for some* $\tau \in \mathrm{Aut}_V(W) \cap \mathrm{Ker}$.

Of course, since $|S| < \infty$, there are only finitely many choices for $\epsilon(\phi, s, t)$ for any ϕ. Thus there are only finitely many cosets of $\mathrm{Aut}_V(W) \cap \mathrm{Ker}$ in Ker, and the proposition will follow from the fact (already shown!) that Ker has finite index in $\mathrm{Aut}(W)$.

Proof. Let $\phi, \phi' \in \mathrm{Ker}$ as in the statement of the lemma. Following [Howlett, *et al.* (1997)], we define $\widehat{\phi}(\alpha_s) = w(\phi, s, s)\alpha_s$ $(= r_{wsw^{-1}}\alpha_s$ for $w = w(\phi, s, s))$; define $\widehat{\phi}'$ similarly.

It is not hard to see that $\{\widehat{\phi}(\alpha_s) \mid s \in S\}$ and $\{\widehat{\phi'}(\alpha_s) \mid s \in S\}$ are both bases for V. Moreover, our assumptions on ϕ and ϕ' imply that

$$\langle \widehat{\phi}(\alpha_s), \widehat{\phi}(\alpha_t) \rangle = \langle \widehat{\phi'}(\alpha_s), \widehat{\phi'}(\alpha_t) \rangle$$

for all $s, t \in S$. That is, there is an orthogonal transformation τ such that $\tau\widehat{\phi}(\alpha_s) = \widehat{\phi'}(\alpha_s)$ for all $s \in S$. It is left as an exercise for the reader to translate this equation involving elements of V into the language of W, finishing the proof of the lemma.   □

As explained above, this completes the proof of Proposition 7.1.   □

7.2.3   The second step

Proposition 7.2 $[\mathrm{Aut}_V(W) : \mathrm{Inn}(W)]$ *is finite.*

This fact is an almost immediate consequence of the following:

Theorem 7.4   *Suppose W is an infinite Coxeter group corresponding to two different irreducible root systems, Φ_1 and Φ_2, with respective simple systems Δ_1 and Δ_2 that span the respective spaces V_1 and V_2. Suppose $\phi : V_1 \to V_2$ gives a bijection between Φ_1 and Φ_2 which preserves the bilinear form $\langle \cdot, \cdot \rangle$. Then for some $w \in W$, $\Delta_2 = \pm\phi(w\Delta_1)$.*

The above theorem say that the corresponding sets of simple roots are essentially "conjugate".

Assuming Theorem 7.4 has been proven, we can finish the proof of Proposition 7.2, and hence of Theorem 7.3. Suppose that $\phi \in \mathrm{Aut}_V(W)$, so that for all $w \in W$, $\phi(w) = \tau w \tau^{-1}$ for some orthogonal transformation $\tau : V \to V$. (Here we consider w as an orthogonal transformation as well.) Theorem 7.4 gives us a permutation π of the simple roots Δ, an element $w \in W$, and a choice of sign, $\epsilon \in \{\pm 1\}$, such that

$$\phi(\alpha) = \epsilon w \pi(\alpha)$$

for all $\alpha \in \Delta$. The bilinear form $\langle \cdot, \cdot \rangle$ is preserved by $\epsilon\pi$, so there is an orthogonal transformation τ' such that $\tau(\alpha) = \epsilon\pi(\alpha)$ for $\alpha \in \Delta$ (namely, $\tau = w\tau'$). Obviously τ' preserves Φ, so there is an automorphism $\phi' \in \mathrm{Aut}_V(W)$ such that

$$\phi'(r_\alpha) = \tau' r_\alpha \tau'^{-1}$$

for $\alpha \in \Delta$. It now follows at once that $\phi \in \text{Inn}(W)\phi'$. There are only finitely many choices of permutations π and signs ϵ, so there can only be finitely many choices for ϕ', so the number of cosets of $\text{Inn}(W)$ in $\text{Aut}_V(W)$ is finite, as desired.

So we must prove Theorem 7.4. The theorem follows from a sequence of lemmas which concerns the structure of the root systems Φ_i. In order to state these lemmas, we must further investigate properties of the *Tits cone* (*cf.* 1.5.2).

7.2.4 *A quick aside: properties of the Tits cone*

Generally speaking, a *cone* in a real vector space V is a subset of V closed under addition and positive scalar multiplication. For instance, consider the following subset of $\mathbb{E}^2$:

$$\{(x, y) \mid y > 0\} \cup \{(0, 0)\}.$$

It is easy to see that this is a cone (see 1.2.3 for a motivation for this example). More generally, any convex (*i.e.*, having interior angle at most π) sector with apex at the origin in $\mathbb{E}^2$ is a cone.

In the dual space V^* ($= \{f : V \to \mathbb{R} \mid f \text{ is linear}\}$), we may define the *dual* K^* to a cone $K \subseteq V$:

$$K^* = \{f \in V^* \mid f(v) \geq 0, \text{ for all } v \in K\}.$$

Let W be a Coxeter group realized as a group of orthogonal transformations on the vector space V, and let Δ be a set of simple roots for W. We let

$$X = \big\{ \sum_{\alpha \in \Delta} \lambda_\alpha \alpha \mid \lambda_\alpha \geq 0, \text{ and } \lambda_\alpha \neq 0 \text{ for some } \alpha \in \Delta \big\}.$$

Clearly X is a cone in V.

It can be shown that for any W, V, and Δ as above, there is a functional $f \in V^*$ such that $f(\alpha) > 0$ for all $\alpha \in \Delta$, so in fact $f(v) > 0$ for all $v \in X$. Denote by C the collection of all such maps f. It is then not hard to see that the closure $\overline{C}$ is the dual X^*.

We use this set C to define the *Tits cone U*:

$$U = \bigcup_{w \in W} w\overline{C},$$

where the action of W (as in $w\overline{C}$) is defined by $(wf)(v) = f(w^{-1}v)$ for all $v \in V$.

Remark 7.3   Note that $\overline{C}$ is a fundamental domain for the Tits cone U, as the Tits cone consists of translates of $\overline{C}$ under the action of W. We also note that $\overline{C}$ can be subdivided naturally into subsets C_T which correspond nicely to the faces in the decomposition of the fundamental chamber C given in 1.5.2.

The Tits cone U plays the role (in general) which was played by the entire vector space $V = V^*$ in case W was finite. Indeed, if W is finite, one can show that $U = V^*$. (This is an "if and only if" proposition: if $U = V^*$, then W must be finite; see Section 5.13 of [Humphreys (1990)] for more details.)

Using the definitions of X, C, and U, we can show the following fact:

Lemma 7.5   *The dual U^* of the Tits cone is given by*

$$\{v \in V \mid w^{-1}v \in \overline{X} \text{ for all } w \in W\}.$$

Proof.   Exercise.   □

In some cases, this enables us to express U^* in terms of the radical of the bilinear form $\langle \cdot, \cdot \rangle$ corresponding to W. (Recall that the radical of the form $\langle \cdot, \cdot \rangle$ is the set of vectors $\{v \in V \mid \langle v, v' \rangle = 0 \text{ for all } v' \in V\}$ which are "orthogonal to everything".)

Proposition 7.3   *Let W be an irreducible affine Euclidean Coxeter group, and let* rad *be the radical of the corresponding bilinear form on the real vector space V. Then* rad *is one-dimensional, and the dual to the Tits cone is given by*

$$U^* = \mathbb{R}_{\geq 0}\rho,$$

where $\mathbb{R}_{\geq 0}$ is the set of non-negative real numbers and ρ is some non-zero vector in the radical rad.

This proposition is proven carefully in [Howlett, *et al.* (1997)]. A similarly useful characterization of U^* can be made in case W is hyperbolic.

Remark 7.4   The reader should take care not to confuse the notion of hyperbolicity used here with *word* hyperbolicity (discussed in Chapter 1). Here, W is said to be *hyperbolic* if it can be realized as a group of orthogonal transformations relative to a symmetric bilinear form $(\cdot, \cdot)$ on the vector space V spanned by $\{v_1, ..., v_n\}$, in such a manner that

$$(x_i, x_j) = \begin{cases} 0 & \text{if } i \neq j, \\ 1 & \text{if } i = j \neq n, \\ -1 & \text{if } i = j = n. \end{cases} \tag{7.8}$$

It is known (see [Humphreys (1990)]) that there are only finitely many hyperbolic Coxeter groups of rank greater than 3 (and there are none of rank greater than 10).

We will not elaborate on the hyperbolic case here; a full treatment of the relevant results can be found in [Howlett, *et al.* (1997)].

Throughout the remainder of the section, we let U_i denote the Tits cone corresponding to the system Φ_i, and let $\overline{C}_i$ be the corresponding "fundamental chamber" in this cone, so that

$$U_i = \bigcup_{w \in W} w\overline{C}_i.$$

7.2.5 The proof of Theorem 7.4

Lemma 7.6 *If W is either an irreducible affine Euclidean Coxeter group or a hyperbolic Coxeter group, then there exist $f \in C_2$ and $\epsilon \in \{\pm 1\}$ such that $\epsilon \phi^*(f) \in U_1^\circ$.*

Here, $\phi^* : V_2^* \to V_1^*$ induced by $\phi : V_1 \to V_2$ in the usual fashion: $(\phi^*(f))(v_2) = f(\phi(v_2))$, and Y° represents the interior of the set Y, for any Y. Lemma 7.6 gives us the toehold we'll need in a little while to prove conjugacy of the simple root systems.

Proof. Let us consider the affine Euclidean case. (The hyperbolic case can be proven in similar fashion.) By Proposition 7.3, we may express the dual Tits cones U_1^* and U_2^* by means of the radicals of the respective bilinear forms:

$$U_i^* = \mathbb{R}_{\geq 0}\rho_i,$$

where ρ_i is a non-zero vector in the radical of V_i. We may replace ϕ by $-\phi$ if needed in order to assume that $\phi(U_1^*) = U_2^*$.

Using properties of the Tits cone derived in the previous subsection, we see that $\phi^*(\overline{U}_2) = \overline{U}_1$. Thus by applying ϕ^* to the interior $(\overline{U}_2)^\circ$, we obtain $(\overline{U}_1)^\circ$, which is in turn U_1°. (Why?)

Now take any $f \in C_2$, the "fundamental chamber" for the Tits cone U_2. Because $C_2 \subseteq U_2^\circ$, $\phi^*(f) \in U_1^\circ$, as desired. □

As in 1.2.2, we denote by Π_i the collection of positive roots in the system Φ_i.

Lemma 7.7   *Theorem 7.4 holds for irreducible affine Euclidean Coxeter groups, and for hyperbolic Coxeter groups.*

Proof.   By replacing ϕ^* with $-\phi^*$ if needed, we may assume (by Lemma 7.6) that $\phi^*(f) \in U_1^\circ$ for some $f \in C_2$. By the definition of U_1, $\phi^*(f) \in w\overline{C_1}$, for some $w \in W$, so $(w^{-1}\phi^*(f))(\alpha) \neq 0$, for all $\alpha \in \Phi_1$. (This follows from the definition of C_2 and our hypotheses on ϕ itself.) From the definition of C_1, $w^{-1}\phi^*(f) \in C_1$.

By "inverting" the definition of C_2, we obtain

$$\Pi_2 = \{\alpha \in \Phi_2 \mid f(\alpha) > 0\},$$

and by applying the action of w and ϕ,

$$\Pi_2 = \{\phi(w\alpha) \mid \alpha \in \Pi_1, (w^{-1}\phi^*(f))(\alpha) > 0\}.$$

At last, since $w^{-1}\phi^*(f) \in C_1$, we can write this last set as $\phi(w\Pi_1)$, so $\Pi_2 = \phi(w\Pi_1)$. From this it follows (how?) that $\Delta_2 = \phi(w\Delta_1)$, as desired.

Note that we must allow ourselves the leeway of "$\pm$" in the statement of Theorem 7.4, since we may have replaced ϕ with $-\phi$. □

Because every Coxeter group of rank 3 is either spherical, affine Euclidean, or hyperbolic (usually the last), Theorem 7.4 is now known to be true for any rank-3 Coxeter group. (Of course, the theorem is trivial in case W is spherical!)

In order to extend the theorem to general Coxeter groups, we turn to the dominance order on the root systems Φ_i (see Exercise 1.21). A *subsystem* $\Phi' \subset \Phi$ is a subset of the roots which is closed under the action of some reflection subgroup W_T of W. The *rank* of such a subsystem is the cardinality of the set T of reflections generating the corresponding subgroup.

Note that if $\Phi_1' \subseteq \Phi_1$ is a subsystem of Φ_1, then $\Phi_2' = \phi(\Phi_1')$ is a subsystem of Φ_2, by our hypotheses on the map ϕ. If the rank of Φ_1' (and therefore of Φ_2') is at most 3, Lemma 7.7 guarantees us $w \in W$ and $\epsilon \in \{\pm 1\}$ such that $\Delta_2' = \epsilon\phi(w\Delta_1')$. (Of course, the sets Δ_i' are the sets of simple roots of the subsystems, corresponding to the generating reflections of the subgroups W_{T_1} and W_{T_2} leaving invariant the respective subsystems.)

Lemma 7.8 *The map ϕ either preserves or reverses the dominance order on any subsystem Φ_1' of Φ_1 of rank at most 3.*

Proof. Let us assume that $\Phi_2' = \phi(\Phi_1')$ yields Δ_2' such that $\Delta_2' = \phi(w_0 \Delta_1')$, for some $w_0 \in W$ (that is, $\epsilon = 1$). We note that the map $\psi(w) = \phi w \phi^{-1}$ is a bijection (in fact, isomorphism!) taking W_{T_1} to W_{T_2} (here, of course, both ϕ and w are considered as linear transformations).

Take $\alpha \succeq \beta$, α and β in Φ_1'. We must show that the set of elements of W_{T_2} "negating" $\phi(\alpha)$ is a subset of those "negating" $\phi(\beta)$. That is,

$$\{w \in W_{T_2} \mid w\phi(\alpha) \in -\Pi_2'\} \subseteq \{w \in W_{T_2} \mid w\phi(\beta) \in -\Pi_2'\}. \qquad (7.9)$$

Making use of the map ψ defined above and that fact that $-\Pi_2' = \phi(-w\Pi_1')$, the lefthand side of (7.9) can be written

$$\phi\{w \in W_{T_1} \mid w_0^{-1}w\alpha \in -\Pi_1'\}\phi^{-1}.$$

Now we use the fact that $\alpha \succeq \beta$ to conclude that this last set is contained in

$$\phi\{w \in W_{T_1} \mid w_0^{-1}w\beta \in -\Pi_1'\}\phi^{-1}.$$

But this is precisely the righthand side of (7.9), and we are done.

In case $\epsilon = -1$ (so that $\Delta_2' = -\phi(w_0 \Delta_1')$), the map ϕ *reverses* the dominance order on Φ_1'. $\qquad\square$

Now to tackle the entire system Φ_1:

Lemma 7.9 *The map ϕ either preserves or reverses the dominance order on all of Φ_1.*

Proof. Given $\alpha \succ \beta$, replace ϕ with $-\phi$ if needed, so that $\phi(\alpha) \succ \phi(\beta)$ holds as well. Let $\gamma, \delta \in \Phi_1$ such that $\gamma \succ \delta$. If $\langle \beta, \gamma \rangle \neq 0$, the subsystem $\Phi_1(\{\alpha, \beta, \gamma\})$ of Φ_1 generated by $\{\alpha, \beta, \gamma\}$ is infinite and irreducible, and of rank at most 3. From Lemma 7.8 and Exercise 1.21, there exists $\zeta \in \Phi_1(\{\alpha, \beta, \gamma\})$ such that $\zeta \succ \gamma$ and $\phi(\zeta) \succ \phi(\gamma)$ both hold. Applying Lemma 7.8 to $\Phi_1(\{\gamma, \delta, \zeta\})$ now allows us to conclude that ϕ preserves the dominance order on this subsystem, so that $\phi(\gamma) \succ \phi(\delta)$, as desired.

So assume that $\langle \beta, \gamma \rangle = 0$. Applying Exercise 1.20, we obtain $\zeta \in \Phi_1$ such that $\langle \beta, \zeta \rangle \neq 0$ and $\langle \gamma, \zeta \rangle \neq 0$. The reader is now asked to construct the appropriate "chain" of roots and subsystems that allows the conclusion that $\phi(\gamma) \succ \phi(\delta)$ (applying Lemma 7.8 and Exercise 1.21 as needed). $\qquad\square$

One more lemma is required:

Lemma 7.10   *Suppose that ϕ preserves the dominance order on Φ_1. Then the set of positive roots $\alpha \in \Pi_1$ such that $\phi(\alpha) \in -\Pi_2$ is finite.*

Proof.   Let P denote the set described in the statement, and suppose that $\gamma \in P$. Select a positive root δ which is minimal (in the dominance order) among all positive roots (as mentioned in Exercise 1.21, the set of such roots is finite; see [Brink and Howlett (1993)]) such that $\gamma \succeq \delta$. Thus $\phi(\gamma) \succeq \phi(\delta)$, so that $\phi(\delta) \in -\Pi_2$, and $\delta \in P$.

Because δ is a "minimal" root, its image under ϕ should be "small" as well. To make this more precise, we define the set

$$Q = \{\alpha \in \Pi_2 \mid -\phi(\beta) \succeq \alpha, \text{ for some minimal } \beta \in P\}.$$

Note that for any $\alpha \in Q$, $r_{-\phi(\beta)}\alpha$ is negative, for some minimal positive root β. As above, there are only finitely many choices of minimal positive roots, and for each such choice of root, the number of positive roots negated by $r_{-\phi(\beta)}$ is finite (see 1.3.2). This implies that Q is finite.

The lemma follows easily from this. $\square$

At last we can finish the proof of Theorem 7.4. Choose $w \in W$ such that

$$P(w) = \{\alpha \in \Pi_1 \mid \phi(w\alpha) \in -\Pi_2\}$$

is as small as possible (Lemma 7.10 implies that it is always finite). Suppose $P(w)$ is not empty, and so contains a simple root $\beta \in \Delta_1$. It is not hard to see that

$$r_\beta(P(w) \setminus \{\beta\})$$

is a set of the above form (exercise!), and obviously it has smaller cardinality. This contradiction implies that $P(w)$ is empty, and thus for some w, $\phi(w\alpha) \in \Pi_2$ for all $\alpha \in \Pi_1$. Theorem 7.4 follows.

7.3   Dehn twists and decompositions: Aut(W) for some large-type groups

Let (W, S) be an arbitrary Coxeter system. As we saw in Chapter 2 (and again in Chapter 6), W can be realized as the result of a series of free products with amalgamation (viewed as the fundamental group of a tree of groups). If this decomposition is nice enough, it is possible to understand

Aut(W) by describing the behavior of an arbitary automorphism on each of the pieces into which W is decomposed.

This is the tack we shall take in this section. The reader is encouraged to compare the methods undertaken here to those used by E. Rips and Z. Sela ([Rips and Sela (1994)]) to compute Aut(G) for certain word hyperbolic groups and by G. Levitt ([Levitt (preprint)]) to compute Aut(W) for generalized Baumslag-Solitar groups, limit groups, and other groups.

7.3.1 Centralizer chasing and visual decompositions: reprise

We make further use of the method of "centralizer chasing" introduced in Chapter 6.

The groups to which we apply this method now are certain even Coxeter groups. Precisely, throughout this section we let (W, S) be a large-type, reflection independent, even Coxeter system whose diagram $\mathcal{V}$ is connected. (W must be rigid; this follows from Theorem 3.8 and Exercise 5.9. The results described in this section will be true after passing to a subgroup of finite index if we omit the words "reflection independent" from the previous sentence.) In fact, such groups can be found by applying Theorem 3.8.

The arguments in Chapter 6 show us that if the diagram $\mathcal{V}$ is "sufficiently well-connected", then W will in fact be strongly rigid, so that its automorphism group is very easy to describe: Aut(W) is a semidirect product of Diag(W) by Inn(W).

Even if $\mathcal{V}$ is not so well-connected, it is often possible to break $\mathcal{V}$ up into subdiagrams which are themselves well-connected, and so on which a given $\alpha \in$ Aut(W) will behave as conjugation by a fixed element of W. (The group W will be isomorphic to a free product of the groups corresponding to these subdiagrams, amalgamating over the junctions at which the subdiagrams are joined.)

What do these subdiagrams look like? That is, what are the maximal parabolic subgroups of W on which any $\alpha \in$ Aut(W) acts as if by conjugation?

Given $\alpha \in$ Aut(W), $(W, \alpha(S))$ is again a Coxeter system, with diagram $\mathcal{V}'$ isomorphic to the original diagram, $\mathcal{V}$. Reflection independence implies that there is some permutation $\pi : S \to S$ such that $\alpha(s)$ and $\pi(s)$ are conjugate to one another. It is not hard (see Exercise 11) to show that $\beta : W \to W$ defined by $\beta(\pi(s)) = s$ is a diagram automorphism, so that by replacing α by $\beta \circ \alpha$, we may assume that s and $\alpha(s)$ are conjugate to one another.

We can now consider chord-free circuits, much as was done in Chapter 6. Exactly as in that chapter, to a chord-free circuit C in $\mathcal{V}$ there corresponds a chord-free circuit C' in $\mathcal{V}'$ of the same length, such that each vertex s of C is conjugate to a unique vertex s' of C'. (In fact, Theorem 5.4 gives the existence of an edge-labeled graph isomorphism which takes C to C'.) Moreover, there is a single group element $w \in W$ satisfying $wsw^{-1} = s'$ for every s on C. That is, $wsw^{-1} = \alpha(s)$ for all s on C.

Continuing to argue as in Chapter 6, suppose that two chord-free circuits, C_1 and C_2, intersect in more than a single vertex or a pair of adjacent vertices. Then the conjugating words w_i for C_i ($i = 1, 2$) are equal. This suggests a means of constructing the subdiagrams described above: begin with any chord-free circuit in $\mathcal{V}$. (Such a circuit will always exist unless $\mathcal{V}$ is itself a chain whose extremal edges are not labeled $2(2k + 1)$, $k \geq 1$.) To the collection of C's vertices add the vertices of any chord-free circuit overlapping C in more than a single vertex or a pair of adjacent vertices. To the resulting collection add the vertices of any chord-free circuits overlapping one of *these* chord-free circuits in such a manner. This inductive procedure constructs the unique maximal subset of S containing C and generating a parabolic subgroup of W on which the given α acts as a conjugation.

The union of the subsets of S constructed in this fashion may not be all of S: there may be vertices $s \in S$ which do not lie on any chord-free circuit. However, it turns out that we may assume such vertices do not exist in $\mathcal{V}$, as long as we are content to compute $\text{Aut}(W)$ up to a subgroup of finite index (see [Bahls (2004b)] for details). For simplicity, let us adopt this easygoing point of view. Therefore, we may assume that $\mathcal{V}$ has no spikes and no bridges (edges which do not lie on any circuit and whose removal separates $\mathcal{V}$). See Figure 7.3.1 for examples: the diagram on the left has bridges and spikes; the one on the right does not.

The picture of $\mathcal{V}$ that we now obtain is a simple one: $\mathcal{V}$ is a union of subdiagrams U_i, the intersection of any two of which is either a single vertex s or an edge $[st]$. Following [Bahls (2004b)], we call each such subdiagram U_i a *unit* of the diagram $\mathcal{V}$. Thus W decomposes visually as a free product of the groups corresponding to the units U_i, with amalgamation over some collection of finite subgroups. Two units U and U' are said to be *adjacent* if their intersection is nonempty.

On each unit, α acts as if by conjugation. Clearly the ratio of conjugating elements corresponding to "adjacent" units is contained in the centralizer of the junction of those units. The following subsection will make this ratio more precise, giving us an explicit description of $\text{Aut}(W)$ and allowing a proof of Theorem 3.16 from Chapter 3.

First, however, we impose one more restriction on the diagram $\mathcal{V}$. For reasons to be explained later, we suppose that $\mathcal{V}$ contains no vertex s such that $\mathcal{V} \setminus \{s\}$ has more than 2 connected components. (We say that $\mathcal{V}$ has *no vertex branching*, or is NVB. This property was illustrated in Figure 3.5.)

The reader should be able to supply a short proof of the following:

Lemma 7.11 *Let (W, S) be a large-type, reflection independent, even Coxeter system whose connected diagram has NVB, no bridges, and no spikes. Then for every unit U (as defined above) in $\mathcal{V}$, the centralizer $C(U)$ is trivial.*

7.3.2 *The ratios of conjugating elements*

Suppose that U_1 and U_2 are units of $\mathcal{V}$ obtained as above, with vertex sets S_1 and S_2, respectively. Let $w_1, w_2 \in W$ such that $\alpha(s_i) = w_i s_i w_i^{-1}$ for all $s_i \in W_{S_i}$, $i = 1, 2$. Let $J = S_1 \cap S_2$, so that $|J| \le 2$, and if $J = \{s, t\}$, $m_{st} < \infty$. As above, $w_1^{-1} w_2 \in C_W(W_J)$.

If $J = \{s, t\}$, then $C_W(W_J) = \{1, (st)^{\frac{m_{st}}{2}}\}$. If $J = \{s\}$, $C(s) = C_W(W_J)$ is more complicated. Theorem 2.5 and the assumption that $\mathcal{V}$ is NVB give us the generating set

$$\{s\} \cup \{u_{s_1 s} \mid s_1 \in S_1\} \cup \{u_{s_2 s} \mid s_2 \in S_2\},$$

for $C(s)$. (Here we have used the notation u_{ts} from Section 2.2.) However, $w_1^{-1} w_2$ will actually be a good deal simpler than this generating set suggests:

Proposition 7.4 *Let (W, S) be a large-type, reflection independent, even Coxeter system whose connected diagram $\mathcal{V}$ has NVB, no bridges, and no spikes. Let U_i, S_i, and w_i be as above, with $S_1 \cap S_2 = \{s\}$. Then $w_1^{-1} w_2$ can be written $u_1 u_2$, where $u_i \in C(s) \cap W_{S_i}$, for $i = 1, 2$.*

Examining our generating set for $C(s)$, this proposition implies $w_2 = w_1 \epsilon_s uv$, where $\epsilon_s \in \{1, s\}$ and u is a product of elements $(ts)^{\frac{m}{2}-1}t$, $t \in \mathcal{V}_1$ (and $m = m_{ts}$) and v is a product of similar elements, taking instead $t \in \mathcal{V}_2$.

In order to prove Proposition 7.4, we need a short lemma.

Lemma 7.12 *Suppose that U and U' are units of $\mathcal{V}$ which both contain the single-vertex junction $\{s\}$ such that $(U \cup U') \setminus \{s\}$ lies in a* *single connected component of* $V \setminus \{s\}$. *Then there is a finite sequence of edges* $[s_1 s], [s_2 s], ..., [s_m s]$ *and corresponding sequence of units* $U = U_1, U_2, ..., U_{m+1} = U'$ *so that* $U_k \cap U_{k+1} = \{s_k, s\}$ *is a junction separating* U_k *and* U_{k+1} *for all* $k = 1, ..., m$.

Proof. There is nothing to prove if $U = U'$, so we assume this is not the case.

Pick and fix vertices $t \in U$ and $t' \in U'$ such that $[st]$ and $[s't]$ are edges in V. Because t and t' lie in the same component of $V \setminus \{s\}$, we can choose a simple path p which does not contain s and which connects t to t'. We assume that p has been chosen as the shortest such path. Concatenating this path with the path $\{[t's], [st]\}$ we obtain a circuit C.

If C is chord-free, then t and t' lie in a common unit, U'', which is separated from U by $\{s, t\}$ and from U' by $\{s, t'\}$, and we are done.

Otherwise, we can "shorten" C to form a chord-free circuit (compare the methods of 6.2.2). Since p was chosen to be the shortest path from t to t' which does not contain s, the only way in which C can fail to be chord-free is if there is some vertex t'' lying on p for which $[st'']$ is an edge in V. In this case, we may divide C into two strictly shorter circuits (one containing $\{s, t, t''\}$ and the other $\{s, t', t''\}$) and induct on the length of the paths into which p has been subdivided to yield the desired conclusion.☐

We now prove Proposition 7.4.

Proof. We know that $w_2 = w_1 w$ for some word $w \in C(\{s\})$. Let us write w as a product

$$w = \epsilon_s \alpha_1 \beta_1 \alpha_2 \beta_2 \cdots \alpha_m \beta_m$$

where $\epsilon \in \{1, s\}$, each α_k is a product of words $(s_l s)^{\frac{m_l}{2}-1} s_l$, $s_l \in S_1$, and each β_k is a product of words $(s_l s)^{\frac{m_l}{2}-1} s_l$, $s_l \in S_2$. We can clearly assume that $\alpha_k \neq 1$ for $2 \leq k \leq m$ and that $\beta_k \neq 1$ for $1 \leq k \leq m-1$.

Let us denote by ϕ the automorphism to which the conjugating words w_1 and w_2 correspond (that is, in particular, $\phi(t) = w_i t w_i^{-1}$ for all $t \in S_i$, $i = 1, 2$). Let ψ denote the map ϕ^{-1}. We know that there exist words w_1' and w_2' such that $\psi(t) = w_i' t w_i'^{-1}$ for all $t \in S_i$, $i = 1, 2$. Furthermore we can write $w_2' = w_1' w'$, where $w' \in C(\{s\})$, so that w' can be written as a product

$$\epsilon_s' \alpha_1' \beta_1' \cdots \alpha_n' \beta_n'$$

for $\epsilon'_s \in \{1, s\}$ and words α'_k and β'_k of forms similar to those of the words α_k and β_k above.

Consider any vertex s_l adjacent to s. If $s_l \in S_1$, Lemma 7.12 shows that there exists a word $\bar{\alpha}_l$ which can be written as a product consisting solely of letters from $S_1 \cup \{s\}$ such that

$$\psi(s_l) = w'_1 \bar{\alpha}_l s_l \bar{\alpha}_l^{-1} {w'_1}^{-1}. \tag{7.10}$$

Similarly, if $s_l \in S_2$, we are guaranteed a word $\bar{\beta}_l$ which can be written as a product consisting solely of letters of $S_2 \cup \{s\}$ such that

$$\psi(s_l) = w'_2 \bar{\beta}_l s_l \bar{\beta}_l^{-1} {w'_2}^{-1}. \tag{7.11}$$

Suppose that t is a vertex in S_1. Then

$$t = \psi \circ \phi(t) = \psi(w_1 t w_1^{-1}) = \psi(w_1) w'_1 t {w'_1}^{-1} \psi(w_1)^{-1}$$

so $\psi(w_1) w'_1 = 1$ since $C(S_1) = \{1\}$ (Lemma 7.11). Now consider $t \in S_2$. Here,

$$t = \psi \circ \phi(t) = \psi(w_1 w t w^{-1} w_1^{-1}) = \psi(w_1)\psi(w)w'_1 w' t {w'}^{-1} {w'_1}^{-1} \psi(w)^{-1} \psi(w_1)^{-1}$$

so $\psi(w_1)\psi(w)w'_1 w' = 1$ since $C(S_2) = \{1\}$. But

$$\psi(w) = \psi(\epsilon_s \alpha_1 \beta_1 \cdots \alpha_m \beta_m) = w'_1 \epsilon_s \tilde{\alpha}_1 w' \tilde{\beta}_1 {w'}^{-1} \tilde{\alpha}_2 w' \tilde{\beta}_2 {w'}^{-1} \cdots \tilde{\alpha}_m w' \tilde{\beta}_m {w'_2}^{-1},$$

where $\psi(\alpha_k) = w'_1 \tilde{\alpha}_k {w'_1}^{-1}$ and $\psi(\beta_k) = w'_2 \tilde{\beta}_k {w'_2}^{-1}$, and equations (7.10) and (7.11) guarantee that $\tilde{\alpha}_k$ can be written as a product consisting solely of letters in $S_1 \cup \{s\}$, and $\tilde{\beta}_k$ can be written as a product consisting solely of letters of $S_2 \cup \{s\}$.

Using $\psi(w_1)w'_1 = \psi(w_1)\psi(w)w'_1 w' = 1$, we obtain

$$\epsilon_s \tilde{\alpha}_1 w' \tilde{\beta}_1 {w'}^{-1} \cdots \tilde{\alpha}_m w' \tilde{\beta}_m = 1. \tag{7.12}$$

Now we expand (7.12) by writing out w':

$$\epsilon_s \tilde{\alpha}_1 (\epsilon'_s \alpha'_1 \cdots \beta'_n) \tilde{\beta}_1 ({\beta'_n}^{-1} \cdots {\alpha'_1}^{-1} \epsilon'_s) \cdots (\epsilon'_s \alpha'_1 \cdots \beta'_n) \tilde{\beta}_m = 1.$$

The occurrences of ϵ'_s can be commuted to the front and multiplied with ϵ_s to yield a single word in $\{1, s\}$. What further reduction can be performed? Assuming w' has been written in reduced form as a product of the words α'_k and β'_k, the only cancellation that can occur is in one of the following subwords:

1. $\tilde{\alpha}_1 \alpha'_1$,
2. ${\alpha'_1}^{-1} \tilde{\alpha}_k \alpha'_1$, $k = 2, ..., m$,

3. $\beta'_n \tilde{\beta}_k \beta'_n{}^{-1}$, $k = 1, ..., m - 1$, or

4. $\beta'_n \tilde{\beta}_m$.

Consider the second case. If $\alpha'_1{}^{-1} \tilde{\alpha}_k \alpha'_1 = 1$, then $\tilde{\alpha}_k = 1$, so

$$\alpha_k = \phi(w'_1) \tilde{\alpha}_k \phi(w'_1)^{-1} = 1.$$

But we have assumed that for $2 \leq k \leq m$, $\alpha_k \neq 1$. Thus the word in the second case above is nontrivial. Similarly we may show that the word in the third case above is nontrivial.

Therefore if $m > 1$, what remains after cancellation is a product in words $(s_l s)^{\frac{m_l}{2} - 1} s_l$ which alternates between blocks such words for $s_l \in A_i$ and blocks of such words for $s_l \in A_j$, and which represents the trivial element. Because the groups that we are considering satisfy the $C'(\frac{1}{6})$ small cancellation condition, any non-trivial word representing the trivial element must contain more than half of a relator appearing in the symmetrized presentation for the group. (See [Lyndon and Schupp (1977)] for more details.) However, this is clearly not the case if $m > 1$. Therefore $m = 1$, and the word w can be written $\epsilon_s uv$ for words u and v described in the statement of the lemma.

Note that we have assumed that w (resp. w') begins with some word α_1 (resp. α'_1) and ends with some word β_m (resp. β'_n); a moment's thought should convince the reader that the other possibilities are similar. ☐

7.3.3 *The unit graph and the structure of an automorphism*

Now that we have some control over the ratios of conjugating elements for adjacent units, we are almost ready to describe the structure an automorphism may possess. We require one more definition.

Let (W, S) be a large-type, reflection independent, even Coxeter system whose connected diagram contains no bridges and no spikes, as in the previous subsections. The *unit graph* $\mathcal{U} = \mathcal{U}(W, S)$ is an unlabeled graph whose vertices are the units of $\mathcal{V}$ and for which there is an edge $[UU']$ if and only if U and U' are adjacent.

Clearly $\mathcal{U}$ is connected. Note that if $s \in S$ is a cut vertex for the diagram $\mathcal{V}$, then the vertices of $\mathcal{U}$ naturally induce two subgraphs, $\mathcal{U}_1(s)$ and $\mathcal{U}_2(s)$, of $\mathcal{U}$, whose vertices consist of those units in one or the other component of $\mathcal{V} \setminus \{s\}$.

We leave it to the reader to prove the following easy result:

Lemma 7.13 *There exists a spanning tree T of the unit graph $\mathcal{U}$ such* *that for every cut vertex $s \in V$, there is a unique edge $[U_1 U_2]$ in T such that $U_i \in \mathcal{U}_i(s)$, $i = 1, 2$.*

We now pick and fix such a tree T. Fix also any vertex U_0 in $\mathcal{U}$ as a basepoint, and assign an orientation to the edge of T in the following manner: if $[U_i U_j]$ is an edge of T which lies on the unique geodesic from U_0 to U_j, then U_i is designated as the initial vertex of $[U_i U_j]$, and U_j as the terminal vertex. (*I.e.*, we orient *away* from U_0.)

We now label each edge $[U_i U_j]$ of T with a label $\phi_{[U_i U_j]}$ chosen in the following manner:

1. If $U_i \cap U_j = \{s, t\}$ is the junction separating U_i from U_j, then $\phi_{[U_i U_j]} \in \{1, (st)^{\frac{m_{st}}{2}}\}$.
2. If $U_i \cap U_j = \{s\}$ is the junction separating U_i from U_j, then $\phi_{[U_i U_j]} = \epsilon u v$, where $\epsilon \in \{1, s\}$, u is a product of words $(s_i s)^{\frac{m_i}{2} - 1} s_i$ for $s_i \in S_i$ and v is a product of words $(s_j s)^{\frac{m_j}{2} - 1} s_j$ for $s_j \in S_j$.

We also label the basepoint U_0 with an arbitrary work $\phi_0 \in W$.

It should now be clear how we will construct an automorphism: given any generator $s \in S$, S lies in some unit U of $\mathcal{U}$, and for such a unit there is a unique geodesic path $\{[U_0 U_1], ..., [U_{n-1} U_n]\}$ in T from U_0 to $U_n = U$. Let ϕ_s be defined by

$$\phi_s = \phi_0 \prod_{i=1}^{n-1} \phi_{[U_i U_{i+1}]}.$$

Then we define the automorphism ϕ itself by

$$\phi(s) = \phi_s s \phi_s^{-1}$$

for all $s \in S$. By considering group relations, it is easy to see that this map is a homomorphism. Moreover, it is not difficult to indicate formulas for the composition and inverse of two maps constructed in the manner described above, proving that ϕ so devised is an automorphism. (We describe composition, leaving it to the reader to compute inverses.)

Let ϕ and ϕ' be homomorphisms defined as above, and let $\psi = \phi' \circ \phi$ denoted their composition. We must describe ψ_0 and $\psi_{[U_i U_j]}$ for all edges of the fixed spanning tree T.

Let $[U_i U_j]$ be an edge in T.

1. If $|U_i \cap U_j| > 1$, define $\psi_{[U_i U_j]} = \phi'_{[U_i U_j]} \phi_{[U_i U_j]}$.

2. If $|U_i \cap U_j| = 1$, $\phi_{[U_i U_j]} = \epsilon uv$, and $\phi'_{[U_i U_j]} = \epsilon' u' v'$, then define $\psi_{[U_i U_j]} = \epsilon\epsilon' uu'v'v$.

3. Lastly, define $\psi_0 = \phi'(\phi_0)\phi'_0$.

Because these automorphisms ϕ are constructed with precisely the flexibility allowed by Proposition 7.4, it is not hard to see that by appropriately choosing ϕ_0, the above construction exhausts all automorphisms ϕ satisfying $\phi(s) \sim s$ for all $s \in S$. Moreover, one may easily show that the choices of ϕ_0 and $\phi_{[U_i U_j]}$ uniquely determine ϕ.

Recall (see 7.3.1) that we may assume $\phi(s) \sim s$ for all $s \in S$, perhaps after precomposing it with a diagram automorphism, so that every element of $\mathrm{Aut}(W)$ can be written as a product of a diagram automorphism and a map ϕ constructed as above. Since $\mathrm{Diag}(W)$ intersects the above collection of automorphisms trivially, this decomposition is unique for every map in $\mathrm{Aut}(W)$. Therefore $\mathrm{Aut}(W)$ can be written as a semidirect product of $\mathrm{Diag}(W)$ by the group G of automorphisms ϕ as above. Theorem 3.16 is now proven.

Remark 7.5   Of course, we have not proven Theorem 3.16 in its fullest generality. For instance, greater care is necessary to consider the cases in which $\mathcal{V}$ contains bridges or spikes. The reader may consult [Bahls (2004b)] for details.

It is clear that in the restricted case we have considered, $\mathrm{Out}(W)$ will be finite if and only if there is some cut vertex $s \in \mathcal{V}$, proving the second assertion of Corollary 3.2. (Why is this?) The other statements in that corollary can be proven by considering the form of $C(\{s\})$, $s \in S$.

We remark that though the computations will indeed be messier, the method outlined in this section should be applicable to Coxeter system more general even than those considered in [Bahls (2004b)].

7.4   Artin groups

We refer the reader to 3.5.3 for the definition of an Artin system and an Artin group.

7.4.1   *Rigidity of Artin groups*

To date, much less is known about Artin groups than is known about Coxeter groups. As indicated in Chapter 3, it is not even known yet whether every Artin group is torsion free. It should not be too surprising, therefore, that less is known concerning rigidity properties satisfied by Artin groups, and that the structure of the automorphism group $\text{Aut}(A)$ (for A an Artin group) is quite poorly understood in general.

However, we can often derive information about an Artin group A from sufficient knowledge of its corresponding Coxeter group, W:

Theorem 7.5 *Let $\mathcal{V}$ be a diagram corresponding either to an Artin system (A, S) or to a Coxeter system (W, S). If (W, S) is reflection rigid up to diagram twisting, then (A, S) is also reflection rigid up to diagram twisting.*

Proof. Let S' be another fundamental generating set for the Artin group A, such that S and S' yield the same reflections, $R = R(S) = R(S')$. Let W and W' denote the Coxeter groups corresponding to these two Artin systems. Because $R(S) = R(S')$, adding any relator s' to the presentation determined by S' is equivalent do adding some relator wsw^{-1} to the presentation determined by S. Similarly, adding any t to the latter presentation corresponds to adding some $w't'w'^{-1}$ to the former. Thus $W \cong W'$.

The result now follows immediately from the assumption that (W, S) is reflection rigid. □

We leave it to the reader to derive Theorem 3.18 from this, and to compile a more complete list of provably reflection rigid Artin groups.

Further, the recent paper [Paris (2004)] provides an Artin analogue of D. Radcliffe's dissertation [Radcliffe (2000)] by proving the following result:

Theorem 7.6 *Given two Artin systems (A_1, S_1) and (A_2, S_2) of finite type, $A_1 \cong A_2$ if and only if the corresponding diagrams are isomorphic.*

7.4.2 *Automorphisms of Artin groups*

Little is known about automorphisms of Artin groups.

J. Crisp has made some headway in understanding $\text{Aut}(A)$ for some Artin groups A. In [Crisp (2004)], he investigates Artin groups with diagrams satisfying conditions similar to those imposed by the author in [Bahls (2004b)]. (In particular, Crisp examines Artin groups with connected, large-type diagrams which have no simple circuits of length 3.) The arguments of [Crisp (2004)] provide a generating set for $\text{Aut}(A)$ in some such cases, and relates this automorphism group to the *abstract commensurator* of A.

In addition to the subgroups $\text{Inn}(A)$ and $\text{Diag}(W)$ which play a major role in the study of Coxeter group automorphisms, the involution $s \mapsto s^{-1}$ for all $s \in S$ takes center stage in studying $\text{Aut}(A)$. (Of course, this involution is trivial in the Coxeter case!)

Crisp's work makes considerable use of the geometry of the *Deligne complex* of a given Artin group, an object closely related to the Davis complex of a Coxeter group. It is known that in a number of cases (including the large-type cases in which Crisp is interested), the Deligne complex admits a CAT(0) metric, allowing the application of a number of fruitful methods whose employment we've already seen.

7.5 Exercises

1. Prove that $(\mathcal{S}, \cdot)$ defined as in Section 7.1 is a groupoid. Furthermore, show that the map $\hat{\beta} \in \text{Aut}(W)$ induced by $\beta \in \text{Aut}(\mathcal{S})$ (as in 7.1.1) is in fact an automorphism.

2. Prove Lemma 7.1 by using the methods of Section 6.2. (What form must $\mathcal{V}$ have if it has only two maximal simplices?)

3. Prove that the generators σ_{sT} defined in 7.1.2 satisfy the relations (7.3), (7.4), (7.5), (7.6), and (7.7).

4. Prove that the statement in Theorem 7.1 beginning "Moreover..." follows from Theorem 7.2.

5. Let P be defined as in 7.2.1. Prove that P has only finitely many orbits under the action of W by conjugation.

6. Give an example which shows that the group $\text{Spe}(W)$ of special automorphisms of W and the subgroup $\text{Ker}(W)$ defined in 7.2.1 need not be the same.

7. Finish the proof of Lemma 7.4.

8. Fill in the details omitted from the proof of Theorem 7.3.

9. Let $C = \{(x, y) \mid y > 0\} \cup \{(0,0)\} \subseteq \mathbb{E}^2$. Prove that the dual cone $C^* \subseteq \mathbb{E}^2$ is the non-negative y-axis. Also, describe the dual cones of the "convex sectors" defined at the beginning of 7.2.4.

10. Prove that $C^{**} = \bar{C}$ for any cone $C \subseteq V$, V a real vector space.

11. Let (W, S) be a large-type, reflection independent, even Coxeter system, and let $\alpha \in \text{Aut}(W)$. Define $\beta : W \to W$ by $\beta(s') = s$ where s' is the unique

element of S conjugate to $\alpha(s)$. Show that β is a diagram automorphism of W.

12. Explain why Lemma 7.13 is necessary, given the way we construct automorphisms of Coxeter groups in Section 7.3.

13. Show that the construction of automorphisms undertaken in 7.3.3 yields all automorphisms ϕ satisfying $\phi(s) \sim s$ for all $s \in S$. Show also that ϕ so constructed is determined uniquely by the choices of ϕ_0 and $\phi_{[U_i U_j]}$ for a fixed spanning tree T (and basepoint U_0) of the unit graph $\mathcal{U}$.

14. Verify the remaining statements of Corollary 3.2 in the restricted case considered in Section 7.3.

15. Use Theorem 7.5 and other results from this text to compile a list of reflection rigid Artin groups.

Bibliography

K. Appel and P. Schupp, "Artin groups and infinite Coxeter groups", *Invent. Math.* **72** (1983) no. 2, 201-220.

P. Bahls, "Even rigidity in Coxeter groups", Ph.D. Dissertation, Vanderbilt University, 2002.

P. Bahls, "A new class of rigid Coxeter groups", *Internat. J. Algebra Comput.* **13** (2003) no. 1, 87-94.

P. Bahls, "Strongly rigid even Coxeter groups", *Top. Proc.*, to appear.

P. Bahls, "Automorphisms of Coxeter groups", *Trans. Amer. Math. Soc.*, to appear.

P. Bahls, "Rigidity of two-dimensional Coxeter groups", preprint.

P. Bahls and M. Mihalik, "Reflection independence in even Coxeter groups", *Geom. Ded.*, to appear.

S. A. Basarab, "Partially commutative Artin-Coxeter groups and their arboreal structure", *J. Pure Appl. Alg.* **176** (2002) 1-25.

C. T. Benson and L. C. Grove, *Finite reflection groups*, Graduate Texts in Mathematics 99, Springer-Verlag, New York, 1996.

N. Bourbaki, *Groupes et algebres de Lie, Chap. IV-VI*, Hermann, Paris, 1981.

N. Brady, J. McCammond, B. Mühlherr, and W. Neumann, "Rigidity of Coxeter groups and Artin groups", *Geom. Ded.* **94** (2002) no. 1, 91-109.

M. Bridson and A. Haefliger, *Metric spaces of non-positive curvature*, Grundlehren der mathematischen Wissenschaften 319, Springer-Verlag, Heidelberg, 1999.

B. Brink, "On centralizers of reflections in Coxeter groups", *Bull. London Math. Soc.* **28** (1996) 465-470.

B. Brink and R. B. Howlett, "A finitness property and an automatic structure for Coxeter groups", *Math. Ann.* **296** (1993) 179-190.

B. Brink and R. B. Howlett, "Normalizers of parabolic subgroups in Coxeter groups", *Invent. Math.* **136** (1999) 323-351.

K. S. Brown, *Buildings*, Springer Monographs in Mathematics, Springer-Verlag, New York, 1989.

K. S. Brown, *Cohomology of groups*, Graduate Texts in Mathematics 87, Springer-Verlag, New York, 1982.

R. Charney and M. Davis, "When is a Coxeter system determined by its Coxeter group?", *J. London Math. Soc.* (2) **61** (2000) 441-461.

H. S. M. Coxeter, "Discrete groups generated by reflections", *Ann. of Math.* **35** (1934) 588-621.

H. S. M. Coxeter, "The complete enumeration of groups of the form $R_i^2 = (R_iR_j)^{k_{ij}} = 1$", *J. London Math. Soc.* **10** (1935) 21-25.

J. Crisp, "Automorphisms and abstract commensurators of 2-dimensional Artin groups", to appear.

M. Davis, "Groups generated by reflections and aspherical manifolds not covered by Euclidean space", *Ann. Math.* **117** (1983), 293-324.

M. Davis, "The cohomology of a Coxeter group with group ring coefficients", *Duke Math. J.* **91** (1998) 297-314.

M. Davis and G. Moussong, "Notes on nonpositively curved polyhedra", in *Low dimensional topology*, K. Bőröczky, Jr., W. Neumann, and A. Stipsicz, eds., Bolyai Society Mathematical Studies 8, Janós Bolyai Mathematical Society, Budapest, 1999.

V. V. Deodhar, "On the root system of a Coxeter group", *Comm. Alg* **10** (1982) 611-630.

W. Dicks and M. J. Dunwoody, *Groups acting on graphs*, Cambridge Studies in Advanced Mathematics, vol. 17, Cambridge University Press, 1989.

A. Dranishnikov and T. Januszkiewicz, "Every Coxeter group acts amenably on a compact space", *Topology Proc.* **24** (1999) Spring, 135-141.

C. Droms, "Isomorphisms of graph groups", *Proc. Amer. Math. Soc.* **100** (1987) no. 3, 407-408.

M. Dyer, "Reflection subgroups of Coxeter systems", *J. Algebra* **135** (1990) no. 1, 57-73.

W. Franzsen, "Automorphisms of Coxeter groups", Ph.D. Dissertation, University of Sydney, 2001.

W. Franzsen and R. B. Howlett, "Automorphisms of Coxeter groups of rank three", *Proc. Amer. Math. Soc* **129** (2001) 2697-2616.

W. Franzsen and R. B. Howlett, "Automorphisms of nearly finite Coxeter groups", **Adv. Geom. 3** (2003) 301-338.

A. Hatcher, *Algebraic topology*, Cambridge University Press, 2002.

R. B. Howlett, P. J. Rowley, D. E. Taylor, "On outer automorphisms of Coxeter groups", *Manuscripta Math.* **93** (1997) 499-513.

R.B. Howlett and B. Mühlherr, "Isomorphisms of Coxeter groups which do not preserve reflections", preprint.

J. Humphreys, *Reflection groups and Coxeter groups*, Cambridge Studies in Advanced Mathematics, vol. 29, Cambridge University Press, 1990.

W. Jaco and P. B. Shalen, "A new decomposition theorem for irreducible sufficiently-large 3-manifolds", *Algebraic and geometric topology (Proc. Sympos. Pure Math., Stanford Univ., Stanford, Calif., 1976), Part 2*, 71-84.

L. D. James, "Maps and hypermaps – operations and symmetry", Ph.D. Thesis, Southampton University, 1985.

I. Kapovich and P. Schupp, "Bounded rank subgroups of Coxeter groups, Artin groups, and one-relator groups with torsion", **Proc. London Math. Soc. (3) 88** (2004) no. 1, 89-113.

I. Kapovich and P. Schupp, "Relative hyperbolicity and Artin groups", preprint.

A. Kaul, "Rigidity for a class of Coxeter groups", Ph.D. Dissertation, Oregon State University, 2000.

K. H. Kim, L. Makar-Limanov, J. Neggers, and F. W. Roush, "Graph algebras", *J. Algebra* **64** (1980) 46-51.

G. Levitt, "Automorphisms of hyperbolic groups and graphs of groups", preprint.

R. C. Lyndon and P. Schupp, *Combinatorial group theory*, Ergebnisse Series, vol. 89, Springer-Verlag, New York, 1977.

M. Mihalik, "The even isomorphism theorem for Coxeter groups", preprint.

M. Mihalik, "Maximal *FA* subgroups of Coxeter groups", preprint.

M. Mihalik and S. Tschantz, "Visual decompositions of Coxeter groups", preprint.

G. Moussong, "Hyperbolic Coxeter groups", Ph.D. Dissertation, The Ohio State University, 1996.

B. Mühlherr, "Automorphisms of graph-universal Coxeter groups", *J. Algebra* **200** (1998) no. 2, 629-649.

B. Mühlherr and R. Weidmann, "Rigidity of skew-angled Coxeter groups", *Adv. Geom.* **2** (2002) no. 4, 391-415.

G. A. Niblo and L. D. Reeves, "The geometry of cube complexes and the complexity of their fundamental groups", *Topology* **37** (1998) no. 3, 621-633.

G. A. Niblo and L. D. Reeves, "Coxeter groups act on CAT(0) cube complexes", *J. Group Theory* **6** (2003) no. 3, 399-413.

A. Yu. Ol'Shanskii, *Geometry of defining relators in groups*, Kluwer Academic Publishers, Dordrecht, 1991.

L. Paris, "Artin groups of spherical type up to isomoprhism", *J. Algebra* **281** (2004) no. 3, 666-678.

D. Radcliffe, "Rigidity of right-angled Coxeter groups", Ph.D. Thesis, University of Wisconsin, Milwaukee, 2000.

R. W. Richardson, "Conjugacy classes of involutions in Coxeter groups", *Bull. Austral. Math. Soc.* **26** (1982) 1-15.

E. Rips and Z. Sela, "Structure and rigidity in hyperbolic groups", *Geom. Funct. Anal.* **4** no. 3 (1994) 337-371.

M. Ronan, *Lecture notes on buildings*, Academic Press, Boston, 1989.

J. Rotman, *An introduction to the theory of groups*, 4th ed., Graduate Texts in Mathematics 148, Springer-Verlag, New York, 1995.

P. Schupp, "Coxeter groups, 2-completion, perimeter reduction, and subgroup separability", preprint.

J.-P. Serre, *Trees*, Springer-Verlag, Berlin, 1980. (Translation of *Arbres, Amalgames*, SL$_2$, Astérisque vol. 46, 1977.)

H. Servatius, "Automorphisms of graph groups", *J. Algebra* **126** (1989) 34-60.

J. Tits, "Le problème des mots dans les groupes de Coxeter", *1969 Symposia Mathematica (INDAM, Rome, 1967/8), Vol. 1*, Academic Press, London (1969) 175-185.

J. Tits, "Sur le groupe des automorphismes des certains groupes de Coxeter", *J. Algebra* **113** (1988) no. 2, 346-357.

E. B. Vinberg, "Discrete linear groups generated by reflections", *Math. USSR Izvestija* **5** (1971) no. 5, 1083-1119.

Index